Robotics: A Very Short Introduction

VERY SHORT INTRODUCTIONS are for anyone wanting a stimulating and accessible way in to a new subject. They are written by experts and have been translated into more than 40 different languages. The series began in 1995 and now covers a wide variety of topics in every discipline. The VSI library contains nearly 400 volumes—a Very Short Introduction to everything from Indian philosophy to psychology and American history—and continues to grow in every subject area.

Very Short Introductions available now:

Alan Winfield

ROBOTICS

A Very Short Introduction

OXFORD
UNIVERSITY PRESS

Great Clarendon Street, Oxford, OX2 6DP,
United Kingdom

Oxford University Press is a department of the University of Oxford.
It furthers the University's objective of excellence in research, scholarship,
and education by publishing worldwide. Oxford is a registered trade mark of
Oxford University Press in the UK and in certain other countries

British Library Cataloguing in Publication Data

Data available

Library of Congress Cataloging in Publication Data

Data available

ISBN 978-0-19-969598-0

Printed in Great Britain by
Ashford Colour Press Ltd, Gosport, Hampshire

Contents

Preface

You may not realize it, but you rely on robots every day. Robots almost certainly built your car and your washing machine. Order something online, and the chances are that your goods will be collected from the warehouse by a robot. The natural gas that powers your central heating is likely to have come from an undersea gas well serviced by robots. Take the driverless electric train between terminal buildings at any number of international airports, and you will be transported in a robot vehicle. And perhaps most surprisingly, the milk on your kitchen table may well have come from cows milked by a robot.

What these robots all have in common is that they are working behind the scenes, by and large out of sight. So, despite their importance, most people are unaware of their key role in the modern world and have never seen a real, working robot 'in the metal'.

But this is changing. A new generation of robots is being designed to interact with humans, face to robot. Unlike their industrial counterparts, these robots will be human-friendly and much smarter; they will have to be in order to be safe. These robots will be capable of providing direct physical support to humans as robot workmates or companions.

A reasonable prediction is that by 2020 many households will have one or more robots, perhaps a driverless car, several cleaning robots, and an educational or entertainment robot. Not many years later and a gardening robot could be taking care of garden weeds and pests. Limited-function robot companions could well become a reality, if not commonplace, by 2025.

Today, if you ask someone to think of a robot, there's a good chance it will be a robot from a movie: perhaps R2D2 or C-3PO from *Star Wars*, Sonny from *I, Robot*, or Disney's *WALL-E*. The problem is that our expectations of what robots are, or should be, draw much more on fiction than on reality. It's been ninety years since Czech playwright Karel Čapek first used the word 'robot' to describe a humanoid automaton in his play *RUR (Rossum's Universal Robots)*, and for many people the word robot remains synonymous with an intelligent mechanical person.

This book is not about the psychology or cultural anthropology of robotics, interesting as those are. I am an engineer and roboticist, so I confine myself firmly to the technology and application of real physical robots. According to the *Oxford English Dictionary*, robotics is the study of the design, application, and use of robots, and that is precisely what this *Very Short Introduction* is about: what robots do and what roboticists do.

This book begins by asking the question 'What is a robot?' In Chapter 1, I introduce a real robot and use it to describe the important parts of all robots: sensing (robot 'eyes' or 'touch'), actuation (robot 'hands', 'arms', or 'legs'), and intelligence (robot 'brains'). This framework serves two purposes. First, it allows us to build a kind of family tree into which we can fit the existing robots I introduce in Chapter 2. Second, it helps us understand both the strengths and limitations of current robots, and why truly intelligent robots remain for the time being only a future possibility.

The last twenty-five years have seen a profound change in the direction of robotics research and development, triggered by the challenge of designing intelligent robots. That change can be described with the words 'biological inspiration'. Illustrating with examples, including robots that get their energy from food and robots modelled on the rat, I introduce bio-inspired robotics in Chapter 3 and what it means for robotics as a whole.

Although the vast majority of robots in the world today are not humanoid, robots made in our likeness hold a special fascination. They are closest to our image of robots as mechanical people. In Chapter 4, I describe work toward humanoid robots as workplace assistants or companions, illustrating this with examples of humanoid robots that range from cartoon-like to realistic. I also outline the problems of robot safety, trustworthiness, and ethics.

In Chapter 5, I describe a number of trends in current robotics research. These include not only new kinds of robots but also radical new ways of designing robots. I introduce swarm robotic systems—multi-robot systems inspired by social insects. Then I outline evolutionary robotics, a new design approach in which robots are artificially evolved using a process akin to Darwinian evolution.

'Where is robotics going?' is a question that receives huge public attention. The final chapter addresses the question of robotic futures by analysing the technical problems that would need to be solved in order to build three 'thought experiment' robot systems: an autonomous planetary scientist, a swarm of medical micro-robots, and a humanoid robot companion.

Like any transformative technology, robotics holds both promise and peril. One of the aims of this book is to equip you the reader with sufficient understanding to reach an informed judgement about not only what robots could realistically be like, but what robots should, perhaps, *not* be like.

Robotics is a large field and I'm acutely aware of significant areas of robotics that are omitted from this short introduction. To fellow roboticists whose areas of work I've passed over briefly, or missed altogether, I offer my apologies. This book is unashamedly a personal view; partly because it's best to write what you know, but more because important questions in robotics are still controversial. Basic questions such as 'What defines a robot?' and 'When can a robot be labelled intelligent?' remain open. Thus, although I hope this book provides a fair representation of the field, the values and judgements I have offered in relation to the various developments in robotics are very firmly my own.

Acknowledgements

I would first like to acknowledge the many scientists and engineers who invented and built the remarkable robots I describe in this book, some of whom I am privileged to know and to work alongside. Without you this book would not have had much to say.

I am grateful to the Engineering and Physical Sciences Research Council (EPSRC) for supporting my public engagement work. That support has provided me with outstanding opportunities to present robotics to a wide-range of people; many of the questions I have tried to answer in this book come from them.

I would like to thank Oxford University Press for their patient support and advice in bringing this book from an idea in an email, to print. I am indebted also to the OUP's anonymous reviewers for insightful comments that really made me think, especially about the word 'intelligent' applied to robots.

I am very grateful to Kerstin Eder, Lilian Edwards, Owen Holland, Ioannis Ieropoulos, Chris Melhuish, Roger Moore, Martin Pearson, and Adam Spiers for checking sections of the text for accuracy. Marina Strinkovsky proof read the final draft and acted as my target readership, for which I am deeply grateful.

List of illustrations

Chapter 1
What is a robot?

Figure 1 shows an e-puck robot. The e-puck is not a real-world robot, nor is it a toy. Designed by Francesco Mondada and his colleagues at EPFL for teaching and research in mobile robotics, the e-puck is a small, wheeled mobile robot and so makes a good candidate for us to look at a robot from the inside. The e-puck is also an open-source robot, which means that all of its plans and software have been made freely available.

Looking at the e-puck, notice that the robot has wheels, driven by electric motors. In this picture we can only see the wheel on the

1. **The e-puck educational mobile robot**

right-hand side, but there's another one on the opposite side. It can drive forwards, backwards, turn, and speed up or slow down. The e-puck robot has a forward-facing camera (similar to the ones in mobile phones), and around its body a total of eight 'proximity sensors'. The proximity sensors use infrared light to sense when objects are very close to the robot's body—within a centimetre or two—so they give the robot something akin to a sense of touch.

Although they are not labelled in Figure 1, the e-puck also has three microphones and a device called an accelerometer which gives the robot something like a sense of balance, rather like your inner ear. For such a small robot, the e-puck has a surprisingly rich sensorium. It can see with its camera, hear with its microphones, sense close objects around it with its proximity sensors, and sense shocks (like sudden collisions) with its accelerometer. The e-puck robot can signal (for instance, to other e-pucks) with a ring of eight lights around its body, or by making sounds with a speaker.

Finally, the e-puck has a single-chip computer, called a microcontroller, on the motherboard underneath the speaker. We can think of the microcontroller as the e-puck's central nervous system. It is connected to each of the motors, input sensors, and signalling devices I have outlined here. The microcontroller, together with its software—its programming—is what makes the robot actually work; it is, in effect, the robot's 'brain'. Figure 2 shows the organization of the e-puck's 'nervous system'.

The natural history of the e-puck

The habitat of the e-puck robot is an artificial one: robotics classrooms and research labs. So what does the e-puck actually do when in its habitat? How does the e-puck respond to stimuli and what are its defining behaviours or traits?

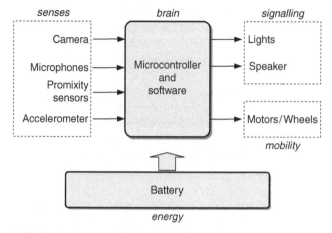

2. The 'nervous system' of the e-puck robot

The first thing that we need to understand about the e-puck robot is that without software *it doesn't do anything at all*. This is a very important observation because it is true of virtually all robots. By software I mean the control program installed in the e-puck's microcontroller 'brain'. Without that control program the e-puck is just an empty shell—a lifeless body without a mind. It is software that animates the robot—bringing it to (artificial) life.

It is also clear that the physical capabilities of the e-puck robot are limited. It's capable only of driving itself around—and even then only on flat or very gently sloping smooth surfaces. Unlike many robots, the e-puck cannot pick up or put down objects—it has no 'gripper' to function like a hand.

If we were to observe e-puck robots in their experimental arena, we might see a group of them in constant motion, staying close but never actually colliding. Rather like a swarm of midges, the e-pucks appear to mill chaotically around each other, yet remain together. A robot naturalist might conclude that these are not solitary creatures but most likely to be found in groups, like social insects.

By carefully observing individual e-pucks within this group, a robot naturalist might deduce they are using their sensors to detect when they are either about to collide or about to leave the group, and their motors to move and turn.

Our robot naturalist might wonder what it is that is making the robots behave in this way. She will observe that the e-pucks are not remotely controlled. While they are busily moving around, no human intervention is needed at all. Like real animals, the robots are behaving *autonomously*. And it is the software in the microcontroller 'brain' that gives the e-pucks their basic behaviours.

Of course, the e-pucks are not completely autonomous. They get their energy from a battery and when the battery runs out the robot simply stops working. So, rather like animals in a zoo, the robots depend on humans to 'feed' them, i.e. change their batteries. This is true for the vast majority of robots—even robots connected to mains power depend on the power grid for energy.

What makes a robot?

In Figure 2, the nervous system of the e-puck robot, I group together the e-puck's *sensors*—the things that it uses to sense the world around it. I also group the things it uses to signal to other robots, as *signalling*, and label its motors as *mobility*. The e-puck's *intelligence* comes from its microcontroller and the software running on it. The e-puck needs a source of *energy*, which comes from its battery. Finally—and perhaps most important of all—the e-puck has a *body*: the transparent plastic chassis we see in Figure 1.

Some or all of those five functions—sensing, signalling, moving, intelligence, and energy, integrated into a body—are present in all robots. The actual sensors, motors, and behaviours designed into a particular robot body shape depend on the job that robot is designed to do. Together they determine what that robot is. Recognizing that all robots are made of the same ingredients gives us a framework for understanding the design of individual robots and how they differ from other robots. The framework will also help us to understand both the strengths and the limitations of current robots.

Table 1 classifies the functions that together constitute a robot, and subdivides each one into various types. All of the robots I introduce in the rest of this book have some combination of the functions in Table 1. For each function type, the table gives both the artificial device (electronic, mechanical, or software) that is typically used to achieve that function, together with the equivalent in animals and humans. A number of the robot function types have no biological equivalent; an obvious example is that no animal gets around on wheels. Similarly, there is no known biological use of radio for communication.

This very unscientific comparison gives the impression that robots can have a bigger range set of senses or signalling than animals, and that is true. However, the balance tips firmly back in favour of animals when we consider that its animal equivalent hugely outclasses almost every part of a current robot. Insects, for instance, have far superior vision, touch, and smell compared with those of robots. Robots are in almost every respect very crude simulacra of animals.

So far I have neglected the biggest part of any robot: its body—the mechanical structure into which all of the other parts are fitted. A robot's shape and body plan, and the positioning of its sensors, wheels (or legs or wings), and arms if it has any, is known as its *morphology*. A robot's morphology has to be very carefully designed so that it is fit for purpose. Unfortunately, the materials and fabrication methods that we have to make robot bodies—normally metal or plastic—are again rudimentary compared with their animal equivalents. Roboticists are often frustrated by these limitations, but with new techniques such as 3D printing we are beginning to see robots with more sophisticated 'organic' body shapes.

Table 1. Robot functions and their approximate natural equivalents

Function	Type	Natural/Biological	Robot/Artificial
Sensing	Vision	Eyes	Cameras
	Hearing	Ears	Microphones
	Orientation (relative)	Vestibular system	Accelerometer
	Orientation (absolute)	Magnetic compass (migrating birds)	Electronic compass
	Olfaction	Nose	Electronic nose
	Short-range sensing	Heat sensors (snake)	Infrared sensors
	Touch (skin)	Skin	Artificial skin
	Touch (whisker)	Whiskers/Skin hair	Artificial whiskers
	Long-range sensing	Echolocation (bat)	Ultrasound/Radar
	Long-range sensing	None	Laser scanner
	Location sensing	None	GPS
Signalling	Lights	Colour (chameleon)	Light bulbs/LEDs
	Sound	Voice	Loudspeaker
	Gesture (facial)	Facial expression	Motorized skin
	Gesture (body)	Body movements	Motors
	Radio	None	WiFi or Bluetooth
Moving	Wheeled	None	Motors and wheels
	Legged	Muscles and legs	Motors and legs
	Flying	Muscles and wings	Motors and wings

Manipulating	Gross Fine	Muscles and arms Hands and fingers	Motors and arms Gripper(s)
Energy	Electrical power	None None None Stomach	Mains power Batteries Solar panels Fuel cells
Intelligence	Controller Behaviours	Brain (matter) Mind (perception and cognition)	Microprocessor(s) Software (control programs)
Body		Bone, cartilage, exoskeleton	Metal and plastic

Despite the limitations I have outlined here, the same basic ingredients are present in animals as in robots. Does this mean that robots are alive? Some roboticists do think of their robots as artificial animals and there is a branch of study that has grown out of Artificial Intelligence and robotics called Artificial Life. But no roboticist would claim their robots to be any more than a simulation of some limited aspects of life or intelligence.

Some robots can certainly demonstrate animated behaviours or responses to stimuli that make them look as if they are alive, but no robots are truly alive. Here are three good reasons why: robots cannot autonomously sustain themselves (i.e. get energy); robots cannot physically grow or repair (heal) themselves; and robots cannot build copies of themselves (reproduce). Only when and if robots could do all of these things would we have to seriously consider the question of whether they are alive. There are exceptions to one of these rules: EcoBot, for instance, a robot that can get energy from digesting food, and which I describe in Chapter 3.

What defines a robot?

Now that we have looked at what makes a robot tick, we can think about a more general definition for the whole robot. Here are three definitions, all useful because they highlight key defining qualities of robots.

A robot is:

1. an artificial device that can *sense* its environment and *purposefully act* on or in that environment;
2. an *embodied* artificial intelligence; or
3. a machine that can *autonomously* carry out useful work.

The first definition is important because it describes a robot in terms of interactions with its working environment: its world. The two key words here are 'sense' and 'act' and they correspond to the robot's sensors, its electronic eyes and ears, and its motors, with which it can manipulate or move in its world. The word 'purposeful' is important too because, without meaningful and in some sense intelligent, actions, the robot would be devoid of purpose or function.

The second definition makes the point that a robot is an Artificial Intelligence (AI) with a physical body. The AI is the thing that provides the robot with its purposefulness of action, its cognition; without the AI, the robot would just be a useless mechanical shell. A robot's body is made of mechanical and electronic parts, including a microcomputer, and the AI made by the software running in the microcomputer. The robot analogue of mind/body is software/hardware. A robot's software—its programming—is the thing that determines how intelligently it behaves, or whether it behaves at all.

We need to be careful here, because behaviour is not just about programming. Behaviour is what happens when a robot interacts with its environment, and that is as much to do with the robot's morphology and environment as with its AI. Roboticists think about a robot's behaviour as something that 'emerges' from the robot's interactions with its operational environment. If, for instance, we are building a walking robot, then, providing it has the necessary hardware (i.e. legs), it is the robot's software that will determine when and why the robot walks forward, stops, or turns. But the actual walking will be a result of the robot's legs, under software control, interacting with the surface it is walking over.

The final definition stresses that a robot should be useful. Robots are suited to jobs that we would regard as dull, dirty, or dangerous. Thus many real-world robots are to be found doing highly repetitive jobs (on factory assembly lines, for instance), in places very hostile to humans (such as the deep sea, or in space), or doing dangerous jobs (such as helping to defuse roadside bombs). It is this definition that perhaps comes closest to the meaning of the Czech word 'robota', meaning literally serf labour or drudgery, which was the original inspiration for the word 'robot'.

Robot autonomy

Another way of thinking about and classifying robots is according to their level of autonomy. Many real-world robots, including some I describe in Chapter 2, are not autonomous but remotely operated by humans. Examples are remotely-operated underwater vehicles (ROVs)—the kinds of robots used for undersea exploration or oil-well maintenance. These are also known as tele-operated robots.

Tele-operated robots only need to have minimal intelligence built into them because the main 'intelligence' controlling the robot is a human. Some tele-operated robots have a greater level of local intelligence and only require the human tele-operator to take control at certain times. For example, unmanned air vehicles (UAVs) might fly themselves (on autopilot) until they reach a preset position, when the human pilot takes control. Such robots are sometimes known as semi-autonomous.

When roboticists talk about autonomous robots they normally mean robots that decide what to do next entirely without human intervention or control. We need to be careful here because they are not talking about true autonomy, in the sense that you or I would regard ourselves as self-determining individuals, but what I would call 'control autonomy'. By control autonomy I mean that the robot can undertake its task, or mission, without human intervention, but that mission is still programmed or commanded by a human.

In fact, there are very few robots in use in the real world that are autonomous even in this limited sense. One rather chilling example of such a robot is the cruise missile: a 'smart weapon' that is able to guide itself to its target without human control. But a human operator is needed to monitor and override the missile's controller if necessary.

It is helpful to think about a spectrum of robot autonomy, from remotely operated at one end (no autonomy) to fully autonomous

Box 1 How intelligent is a robot vacuum cleaner?

A robot vacuum cleaner typically needs five preprogrammed behaviours:

- Sensing and avoiding obstacles, such as furniture or walls;
- Sensing when its battery is about to run out, then 'homing' to its recharging station to plug itself in and recharge;
- Sensing edges under the front of the robot and retreating from those edges so that it doesn't fall over them;
- Sensing that its dust collector is full, then stopping and signalling (to a human) that it needs emptying;
- Or, if none of the above, going forward (the default behaviour).

What may seem odd is that this set of behaviours doesn't include 'cleaning carpet'. In fact, whenever the robot is moving, it is automatically cleaning whatever carpet is under it, so the main carpet-cleaning behaviour is 'go forward'. Of course, all rooms have walls and most have furniture, so the robot can't go forward for long before it will sense an obstacle, which triggers the first behaviour above.

The robot will have a number of proximity sensors and the 'avoiding obstacles' behaviour will be slightly different depending on exactly which sensor is triggered. For instance, if the front-left sensor is triggered, the robot will turn right; similarly, something blocking the front-right sensor will cause the robot to turn left. If both front-left and front-right sensors are triggered at the same time, then the robot must be about to hit something head-on and so should stop and turn away. If the robot is circular, it can simply rotate by some angle bigger than 90 degrees. If that angle is different every time (i.e. the robot chooses—within limits—some random turn), then it means that every time the robot has to avoid something it will turn and set off in a new direction across the room. In this way the robot will eventually cover and vacuum the whole of the room.

Box 1 Continued

This may seem like a very inefficient approach, and indeed it is. The movement pattern of the robot vacuum cleaner is what roboticists call a random walk. It means that some parts of the carpet will be vacuumed several times over, whereas other parts only once (or if you're very unlucky, not at all). However, the likelihood is that after a while the robot will have covered the whole carpet. A more efficient approach, one in which the robot remembered which parts of the carpet it had already cleaned, or makes a 'map' of the room so that it can plan a way of cleaning from one end to the other, would require much more complex and expensive sensors and more 'intelligence', which would mean that the robot would costs thousands rather than hundreds of pounds. The low-intelligence approach, with just five preprogrammed behaviours, is enough to get the job done.

at the other. We can then place robots on this spectrum according to their degree of autonomy.

I should say something here about the difference between autonomous and automatic. Even though the two words actually have similar meanings, roboticists prefer the word autonomous since it gives a sense of more than automatic, and striving for greater autonomy is an important goal in robotics.

We take automatic to mean 'running through a fixed preprogrammed sequence of actions'. A robot arm used to spray-paint cars should therefore be described as automatic and would fall at the low end of a scale of autonomy. But consider a robot vacuum cleaner. Even though this is a relatively simple robot, it would rate a high score for autonomy. The reason is that its actions are determined by its sensory inputs, rather than where it is in a preprogrammed sequence. Put an obstacle in front of the robot and it will take avoiding action. No obstacle, and the robot will keep going ahead.

On a scale of autonomy, a robot that can react *on its own* in response to its sensors is highly autonomous. A robot that cannot react, perhaps because it doesn't have any sensors, is not. It is also important to note that autonomy and intelligence are not the same thing. A robot can be autonomous but not very smart, like a robot vacuum cleaner.

Robot intelligence

I am often asked: 'How intelligent are robots?'—perhaps out of a sense of unease that robots shouldn't be too intelligent. At one level it's a difficult question to answer properly. Making meaningful comparisons between the intelligence of a particular robot and a particular animal is fraught with difficulty, not least because we only have a vague notion of how intelligent most animals are compared with each other and with us humans.

We have a general sense that a crocodile is smarter than an ant, and a cat smarter than a crocodile, but this notion is wrong. Intelligence is not a single measurable quality (like an IQ test) that different animals, or indeed robots, have more or less of. But setting aside this objection, where would we assess the intelligence of, say, a robot vacuum cleaner to be? If pressed, I would suggest right at the bottom.

A robot vacuum cleaner has a small number of preprogrammed (i.e. instinctive) behaviours and is not capable of any kind of learning (see Box 1 for a description). These are characteristics we would associate with very simple animals. So is our robot vacuum cleaner about as smart as a single-celled animal such as an amoeba?

When roboticists describe a robot as intelligent, what they mean is 'a robot that behaves, in some limited sense, *as if* it were intelligent'. The words *as if* are important here. Few roboticists would claim a robot to be truly intelligent. They might claim a robot deserves to be called intelligent, in this qualified sense, because the robot is able to determine what actions it needs to take in order to behave appropriately in response to external stimuli.

My own view is that a robot can be regarded as demonstrating a limited simulation of intelligence while, at the same time, being regarded as intelligent. By analogy, an aeroplane can be considered a simulation of flight, but no one would doubt it is also truly flying. Thus I believe it is hard to argue that a robot that behaves as if it is intelligent is not, within in the very limited scope of those behaviours, properly intelligent. I accept, however, that this view may be controversial. The question of what properties constitute intelligence remains a difficult scientific and philosophical question.

Let's think about robot intelligence a different way. I could have answered the question 'How intelligent are robots?' with the response: 'as intelligent as they need to be to do the job they are designed for'. A robot vacuum cleaner doesn't need to be very smart in order to be good at cleaning carpets, and the same is true for very many robots. But what if the job that we need a robot to do *does* demand much more intelligence? How can we make robots smarter, and what are the current limits on robot intelligence?

There are basically two ways in which we can make a robot behave as if it is more intelligent:

1. preprogram a larger number of (instinctive) behaviours; and/or
2. design the robot so that it can learn and therefore develop and grow its own intelligence.

The first of these approaches is fine, providing that we know everything there is to know about what the robot must do and all of the situations it will have to respond to while it is working. Typically we can only do this if we design both the robot and its operational environment. In this way we can limit the number of unexpected things that might happen, and therefore anticipate and preprogram exactly how the robot must react to each of these events. Factory robots are a good example here—they work in a carefully engineered environment in which the work they have to do is presented in exactly the same position and orientation every time (for instance, welding parts of a car).

But what if we want a robot to be able to work in unstructured environments—anywhere that was not designed with the robot in mind? In this sense almost any outdoor or indoor environment is unstructured. Even offices, which to us seem highly structured, are a problem for robots because of the people in them. Making robots both smart and safe in human environments is a major unsolved problem for roboticists.

For unstructured environments, the first approach to robot intelligence above is infeasible simply because it's impossible to anticipate every possible situation a robot might encounter, especially if it has to interact with humans. The only solution is to design a robot so that it can learn, either from its own experience or from humans or other robots, and therefore adapt and develop its own intelligence: in effect, grow its behavioural repertoire to be able to respond appropriately to more and more situations.

This brings us to the subject of learning robots—something I shall return to in a little more detail later in this book. Suffice it to say robot learning or, more generally, 'machine learning'—a branch of AI—has proven to be very much harder than was expected in the early days of Artificial Intelligence. Thus, although there are plenty of examples of research robots that demonstrate simple learning, such as learning to find their way out of a maze, none has so far demonstrated what we might call general problem-solving intelligence. This is the ability to learn either individually (by trial and error) or socially (by watching and learning from a human teacher or another robot), then generalize that learned knowledge and apply it to new situations. It is the kind of learning that comes naturally to human children.

A very brief history of robotics

Although the words 'robot' and 'robotics', and the science that followed the fiction, are decidedly 20th century, robotics has a long pre-history of ideas and inventions. Perhaps the first known reference to the idea of an 'intelligent' tool that could replace human labour comes from Aristotle, who wrote in 320 BC that 'if every tool, when ordered, or even of its own accord, could do the work that befits it . . . then there would be no need either of apprentices for the master workers or of slaves for the lords'.

The practice of constructing mechanical automata dates back at least 2,000 years. Hero of Alexandria constructed a number of automata, including, in about 60 AD, a self-powered three-wheeled cart. The cart was powered by a falling weight that pulled strings wrapped around its axles, and has recently been discovered to be programmable by means of pegs in the axles, so that the direction of winding of the string on the axle can be reversed. Thus the cart can be programmed to turn and follow a preset route.

The earliest reference to the idea of a humanoid automaton is to be found in Jewish folklore with the Golem: an animated humanoid being made of inanimate matter, normally clay, brought to life by magic. Although its true origins are controversial, one of the most famous stories relates how the 16th-century Rabbi Loew of Prague created a Golem to defend the city's ghetto from attack. A particularly interesting aspect of the mythical Golem is that it would interpret commands literally, with unintended and sometimes disastrous consequences. This is also a property of modern robots, as we roboticists are often painfully reminded.

One of the most interesting and characteristically far-sighted automata of the Renaissance period is Leonardo da Vinci's autonomous cart. Recent research has established this as a working self-powered automaton of remarkable sophistication. Leonardo's cart is powered by clockwork springs, but most notable is the system of replaceable cams controlling both the steering and speed. These allow the cart to be programmed to follow a preset route, starting, stopping, and turning as required. The robot could also be programmed to trigger a special effect at a preset time, such as opening a door on a sculpture mounted on the cart. Leonardo might also be credited with the first design, c.1495, for a complete humanoid robot, with his robot knight. Based on biomechanical principles from his anatomical research, Leonardo's knight had cable-driven arms, head, and jaw.

During the 18th century, mechanical automata reached a high degree of sophistication (as well as fakery). Perhaps the best-known example is French inventor De Vaucanson's 1739 mechanical duck, which famously defecated after 'eating' grain. This was an illusion—the duck actually excreted a premixed preparation of dyed green breadcrumbs. (True robotic artificial digestion would not appear for another 270 years, as I describe in Chapter 3.) More significant in my view is de Vaucanson's flute-playing humanoid automaton, notable for the sophistication and bio-mimicry of its mechanisms for controlling the airflow, via the robot's lip and tongue movements, needed to successfully play the flute.

The subject of this book, modern robotics, depends on several key technologies, including the electric motor to provide robots with actuation, and electronic devices to provide the means to control and automate robots. But new ideas were needed too, and a brilliant group of men in the mid 20th century laid the foundations of modern robotics in what was then (and is sometimes still) called cybernetics.

Notable members included British scientists Alan Turing and W. Ross Ashby, and in the United States Norbert Weiner and Warren McCulloch. Among that group was the brilliant neurophysiologist W. Grey Walter. Walter holds a special place in the modern history of robotics because of his robot 'tortoises', now widely regarded as the first autonomous electronic mobile robots (see Figure 3).

Walter developed his tortoises (so named because they 'taught us', after Lewis Carroll) to demonstrate his ideas on brain function. He was convinced that the connections within the brain, and the number of connections, are of much greater importance in giving rise to intelligence than the number of brain cells. He designed the control system of the robots with, as he put it, 'a simple two cell nervous system' to show that even a very simple system like this can have rich behaviours. The 'cells' were vacuum tubes (the 1940s equivalent of the

3. Replicas of Grey Walter's robot tortoises

transistor) and by ingeniously interconnecting the cells, and the robot's sensors and motors, the robots demonstrated four distinct behaviours.

In its default state, the front wheel of the robot, which provides both steering and traction (like a child's tricycle) rotates continuously, providing the robot with a characteristic cyclic motion; this is the 'explore' behaviour. A photocell mounted at the top of the steering/drive wheel provides the robot with light sensitivity, so it is attracted by a strong light source at some distance, and repelled by the light source when it is very close. The fourth behaviour is obstacle avoidance: the robot's perspex shell provides it with a bump sensor that, when triggered, changes the robot's motion so that it can escape the obstacle.

Walter's robots demonstrated ideas in robotics that to some extent were lost and rediscovered in the 1980s. Neither preprogrammed nor remotely controlled, Walter's tortoises behave in a complex and unpredictable way that is as much to do with the robot's operating environment as with the robot itself. In a fascinating series of experiments, Walter fixed candles to the two tortoises, named Elmer and Elsie, then captured their movements on a long-exposure camera. Elsie and Elmer demonstrated unexpected behaviours. For instance, when placed in a dark room with no other light source, Elsie would catch sight of Elmer, and vice versa, and the two robots would approach each other, then engage in a kind of 'dance'.

These experiments anticipated ideas of behaviour-based robotics, which I describe in Chapter 3, and swarm robotics, which I introduce in Chapter 5. In another sense, Grey Walter's robot tortoises continued a tradition, already established in 19th-century automata, of machines that provide simple models of life, to both educate and entertain.

Chapter 2
What robots do now

Assembly-line robots: robots for making things

Assembly-line robots are the workhorses of robotics: the
unglamorous robots that have been busy in factories for decades.
The usual term for these robots is robot arms, or more technically,
multi-axis manipulators. Figure 4 shows several examples of
industrial robot arms. Each consists of a base that is fixed to the
ground, and a series of segments each connected to the next by a
joint and a motor. At the end of the final segment is a wrist joint
which usually allows any one of a number of special purpose tools
to be attached. These are known as end-effectors.

It is the end-effectors that do the work of the robot. The joint
motors allow individual joints to be moved in different axes, which
means that a combination of movements across the multiple axes
will allow the end-effector to be positioned at more or less any
angle or place within the reach of the robot.

The great strength of industrial robot arms is, in fact, their
great strength — combined with high precision. Typically the
end-effector can be repeatedly positioned with an accuracy of better
than 1mm, which is remarkable given that the robot might weigh
several hundred kilograms. Robot arms on an assembly line are
typically programmed to go through a fixed sequence of moves over

4. Industrial assembly-line robots

and over again, for instance spot-welding car body panels, or spray-painting the complete car. These robots are therefore not intelligent. In fact, they often have no exteroceptive sensors at all.

(They will, however, normally have internal sensors for measuring the angles of each joint.) Thus a spray-painting robot may not be able to sense whether or not there is a car for it to paint. It will rely on the car, or more generally the work, to be positioned in exactly the right place, at the right time, ready for it to begin its work cycle. Thus when we see an assembly line with multiple robot arms positioned on either side along the line, we need to understand that the robots are part of an integrated automated manufacturing system, in which each robot and the line itself have to be carefully programmed in order to coordinate and choreograph the whole operation.

An important characteristic of assembly-line robots is that they require the working environment to be designed for and around them, i.e. a structured environment. They also need that working environment to be absolutely predictable and repeatable. Contrast this with human environments that are typically (from a robot's point of view) unstructured and unpredictable. For this reason, and the fact that assembly-line robots are dangerous to humans because they are both strong and unable to sense people, humans need to be kept at a safe distance. Although a very successful technology, assembly-line robots thus characterize the first wave of robots: robots not designed to interact with humans.

How to program a robot arm

Think of the ease with which you and I pick up and lift a cup of coffee to our lips. It's something that seems to require no conscious effort at all, save the minimal impulse to take a sip of coffee. Yet when we break down the sequence of actions, we see that what's actually happening is far from simple. First, you reach for the cup: an action that almost certainly involves moving both your shoulder and elbow joints. But sensing is also involved, because those movements need to be guided by your vision to where the cup is; again the process is subconscious and you don't even need to be looking directly at the cup for it to work. Once your hand reaches the cup handle, there is a rather complicated set of movements of your finger joints to grasp the handle. Even if you've never before encountered that particular type of coffee cup handle, you still seem to be able to automatically, and without conscious effort, adapt your fingers and their grip to suit its shape.

Then you start to lift the cup. At this point something else remarkable happens: you automatically (and again with no conscious awareness) sense the weight of the cup and the liquid in it and apply just the right amount of force to your shoulder, elbow, and wrist joints. Too little and you wouldn't lift the cup at all; too much and the coffee will be spilled. Then, as you bring the cup toward your lips, you smoothly adjust your wrist so that as your elbow flexes, the cup—and the coffee in it—stays level. This action is, again, without conscious thought, guided by both your vision and the feedback from your arm and the pressure on your fingers. Finally, as you bring the cup into contact with your lips, perhaps tentatively because you sense the coffee is hot, the precise positioning of the cup and your careful tilting of the cup to drink will rely also on the touch sensing of your lips.

What this illustration of a seemingly mundane human action shows is that there is a great deal of coordinated sensing and control of multiple joints going on here. The neural 'programming' that gives you this skill was probably learned during your early development as a baby and toddler. But for a robot arm of the type we see in assembly lines, there is no such process of learning a new skill.

Robot arms either need to be painstakingly programmed, so that the precise movement required of each joint is worked out and coded into a set of instructions for the robot arm or, more often (and rather more easily), 'taught' by a human using a control pad to move its end-effector (hand) to the required positions in the robot's workspace. The robot then memorizes the set of joint movements so that they can be replayed (over and over again). The human operator teaching the robot controls the trajectory, i.e. the path the robot arm's end-effector follows as it moves through its 3D workspace, and a set of mathematical equations called the 'inverse kinematics' converts the trajectory into a set of individual joint movements.

Using this approach, it is relatively easy to teach a robot arm to pick up an object and move it smoothly to somewhere else in its workspace while keeping the object level (like the cup of coffee in our human example). However, unlike our example, most real-world robot arms are unable to sense the weight of the object and automatically adjust accordingly. They are simply designed with stiff enough joints and strong enough motors that, whatever the weight of the object (providing it's within the robot's design limits), it can be lifted, moved, and placed with equal precision.

Similarly, the stiffness and precision of movement, under programmed control, mean that the robot arm can keep an object level without requiring the kind of sensing and constant adaptive control humans need to keep an object level while moving it. It is the stiffness of robot joints that makes them relatively easy to program, and to control precisely and repeatably. In this regard, conventional robot arms could not be more different from human arms or animal limbs.

The robot arm and gripper are a foundational technology in robotics. Not only are they extremely important as an industrial assembly-line robot, but they have become a 'component' in many areas of robotics. Variants of the robot arm can be found attached to mobile platforms such as underwater remotely operated vehicles, bomb disposal robots, or Mars rovers. With widely varying degrees of sophistication, robot arms are essential features of humanoid robots, and I shall turn to the subject of humanoid robot arms, including compliant robot arms, in Chapter 4.

Portering robots: robots that fetch and carry

When we place an order with an online vendor, there's a good chance that our order will be collected from its shelf in the warehouse by a robot, then brought to a central point where human packers put it in a box ready to send to us. Unlike assembly-line robots, warehouse robots are *mobile*. They are portering robots—designed essentially to fetch and carry. However, warehouse robots have one thing in common with assembly-line robots: they work in highly structured environments where humans are, by and large, kept out. Some warehouses run lights-out because the robots don't need lights and humans never need to go into the main storage part of the warehouse (except, of course, if something goes wrong). Warehouse robots typically move along predefined routes through the warehouse.

However, like assembly-line robots, portering robots are highly sophisticated machines. And—also as with assembly-line robots—it is a mistake to judge a single robot in isolation. Portering robots often work in groups called multi-robot systems that in turn form the visible part of an integrated system of great complexity. I shall have more to say about multi-robot systems in Chapter 5.

Warehouse or factory portering robots belong to an important class of robots called automated guided vehicles (AGVs). An AGV is a mobile robot that is able to autonomously navigate from one point to another, often by following a buried wire or markers on the floor. A portering AGV must be able to carry or tow a load, perhaps lifting it and setting it down automatically. A good example is a forklift AGV (in effect a driverless forklift truck): it must be able to sense obstacles in its path, including people, and safely come to a stop. An AGV is usually an electric vehicle, powered by batteries, and so it must also be able to sense its own battery level and determine when to stop working and go to a recharging station.

We can see therefore that a portering AGV needs at least three types of sensors: to detect navigational markers, to detect obstacles, and to sense its own battery level. It also needs to communicate, via radio, with the warehouse or factory system controlling and coordinating the AGVs and, if the AGV has to be loaded or unloaded manually, it will need a simple human–robot interface so that the human operator can tell the AGV when they've finished loading or unloading. All of these requirements add up to a mobile robot that needs to be a good deal more 'intelligent' than its fixed counterpart, the assembly-line robot. It also needs to be strong, reliable, and very safe.

Automated guided vehicles have been in use for many decades, but a newer generation of AGVs is finding application in very human environments. As an example, consider the Aethon TUG:

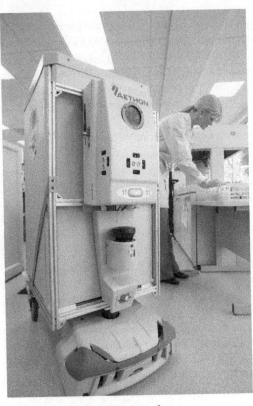

5. The Aethon TUG hospital portering robot

a hospital portering robot, shown in Figure 5. At the time of writing, more than a hundred of these robots are in use in hospitals in the USA, typically for fetching medical supplies, including blood or pharmaceuticals, or delivering samples for analysis. The design of the robot is interesting: it consists of a container on wheels—with compartments or drawers designed for whatever it has to carry—which is hauled by a small 'tug' robot (hence its name).

Unlike its industrial counterparts, the TUG doesn't require buried wires or floor markers to navigate. Each robot is preprogrammed with an electronic 3D map of the hospital, marked with key positions such as the pharmacy, pathology lab, nursing stations, doors, elevators, and, importantly, the robot's recharging station—its 'home' location.

The robot navigates by matching what it 'sees' with its laser range finders, with what it expects to see from its current position and orientation according to its internal map. Think of the laser range finders as sending out fingers of light to measure the distance of almost every point on every object in front of the robot—the robot in effect 'sees' in 3D, with depth measurements. The robot is thus able to track its own position in its internal map. If it senses that the way ahead is clear, the robot will find its way to its destination autonomously; if it senses, for instance, that it has come to the end of a corridor with a T-junction, the robot will consult its internal map in order to decide whether to turn left or right.

The hospital does, of course, require some special infrastructure to support the robot porters: doors and elevators must be fitted with remote controls that allow the robot to 'ask' the door to open, or fetch the elevator and request the floor it wants to go to. And a wireless network (which a hospital is likely to have anyway) is needed so that hospital workers can call for a robot.

Unlike warehouse AGVs, the TUG is required to interact with people; it must operate safely in one of the most demanding of human environments. Basic safety features are, first, that the TUG moves rather slowly and gives audible beeps while it is moving. Second, the TUG is able to sense possible collisions with people or objects, and will come to a stop; it will only start moving again when the way is clear. Human–robot interaction is simple, but effective. Hospital staff can request pickups of deliveries via the WiFi network, and when the TUG arrives at its destination the robot will make a verbal request by 'speaking' a synthesized phrase, to indicate to the human user that it has come to either collect or drop off an item. The TUG has no speech input: just two large buttons which the user presses ('pause' and 'go') after loading or unloading to send the robot on its way. Between deliveries, the TUG will find its way back to its recharging station so that it is ready for work the next time it is required.

Tele-operated robots: keeping the human in the loop

Tele-operated robots (although not necessarily named as such) have, in recent years, come to prominence in a number of important spheres of application. Undersea remotely operated vehicles (ROVs) are now a routine technology for deep-sea exploration, salvage, and maintenance. In the field of military aerospace, tele-operated unmanned air vehicles (UAVs) have since 1990 assumed considerable importance in aerial reconnaissance. Planetary exploration has had remarkable success with the tele-operated deep space probes Pioneer and Voyager, and more recently the Mars planetary rovers Spirit and Opportunity. Tele-robotic devices are finding acceptance by surgeons and are already regarded as key surgical tools for certain procedures. I shall outline examples from each of these real-world applications, but first let us define tele-operated robots.

Tele-operated robotics, or tele-robotics for short, describes the class of robotic devices that are remotely operated by human beings. Tele-operated robots thus contrast with autonomous robots that require no human intervention to complete the given task. The distinction between autonomous and tele-operated robots is blurred, however, since some tele-operated robots may have a considerable degree of local autonomy. The robot might, for example, accept commands such as 'Move forward 30 cm,' and automatically work out and perform the control of its motors and wheels to carry out that command, freeing the human operator from these low-level control operations.

Tele-operated robots share almost all of the characteristics outlined in Chapter 1. They are often equipped with a rich set of sensors, including cameras, and a range of complex actuators, for both moving and manipulating objects. But when we consider tele-operated robots, it is important to consider the whole system, including the human operator, rather than just the robot.

The three main elements that define a tele-operated robot system are the operator interface, the communications link, and the robot itself. Let us consider these in turn.

The operator interface. This will generally consist of one or more displays for the video from the robot's onboard camera(s) and other sensor or status information. In addition, the interface will require input devices to allow the operator to enter commands (via a keyboard), or execute manual control of the robot (via a joystick, for instance).

The communications link. This might utilize a wired connection for fixed tele-operated robots, or wireless for mobile robots. In either case, the communications link will need to be two-way so that commands can be transferred from the operator interface to the robot, and at the same time vision, sensor, and status information can be conveyed back from the robot to the operator. Often, the communications channel will require a high capacity to carry the real-time video from the robot back to the operator.

The robot. The robot's design will vary enormously over different applications and operating environments. But whether the robot rolls on tracks, flies, or swims, it will always need a number of common subsystems. The robot will need electronics for communication, computation, and control. It will need software to interpret commands from the operator interface and translate these into signals for motors or actuators. The software will need to monitor the robot's 'vital signs', including battery levels, and send status information back to the operator alongside data from the robot's cameras and other sensors.

The defining characteristic of tele-operated systems is that the human operator is an integral part of the overall control loop. Video from the robot's onboard camera is conveyed, via the communications link, to the human operator. She or he interprets the scenario displayed and enters appropriate control commands that are transmitted, via the communications link, back to the robot. The robot then acts upon (or moves in) its environment in accordance with the control demands, and the outcome of these actions is reflected in the updated video data to the operator. Hence the control loop is closed.

Anyone who has tried to operate a robot remotely while peering at a screen showing the view from its camera knows how difficult it is, and good design in the operator interface is therefore very important to the success of tele-operated robot systems. A particular design problem is, for instance, how to provide the human operator with an immersive experience of the robot's sense data (i.e. as if they were there). One example of a way to improve the sense of being there is through remote tele-haptics—in other words, allowing the human operator to feel what the robot's touch sensors are touching. A well designed tele-operated robot system acts as a kind of human sense and tool extender, to allow human exploration, inspection, or intervention in environments far too dangerous or inaccessible for humans.

From a robot design point of view, the huge advantage of tele-operated robots is that the human in the loop provides the robot's 'intelligence'. One of the most difficult problems in robotics—the design of the robot's artificial intelligence—is therefore solved, so it's not surprising that so many real-world robots are tele-operated. The fact that tele-operated robots alleviate the problem of AI design should not fool us into making the mistake of thinking that tele-operated robots are not sophisticated—they are.

Let's now look at some representative types of tele-operated robots.

Undersea robots: ROVs

Remotely operated vehicles (ROVs) are the workhorse robots of the offshore oil exploration and drilling industry; they are also essential robots for deep-sea exploration and science. Science ROVs famously played a key role in the exploration of the rich flora and fauna around the deep-sea smoking hydrothermal vents. They are essentially remotely controlled unmanned submarines, often with robot arms with grippers or manipulators so that the ROV operator can undertake maintenance or repair, or, for science ROVs, collect samples. Often ROVs are attached to the mother ship via a cable tether, which provides both power and control signals to the ROV, and the video camera feed(s) and other instrument data from the ROV back to the shipboard human operator.

Military robots

It is often said that robots are most suitable for jobs that are dull, dirty, or dangerous. Firmly in the dangerous category are bomb disposal robots. Pioneered by the British army in the 1970s with the invention of the so-called 'Wheelbarrow', these are remotely operated vehicles whose primary function is to allow bomb disposal experts to inspect vehicles or buildings suspected of having bombs planted inside them without placing themselves at risk.

Bomb disposal robots are essentially rugged radio-controlled mobile robots, usually running on tracks, fitted with a robot arm. Often the robot is fitted with two cameras, one on the body of the robot and another fitted to the end of the robot arm. The latter is especially important as it allows the bomb disposal expert to be able to remotely manipulate the robot arm in order to position the camera, so that they can look, via the camera, for anything suspicious. In addition to the camera(s), the robot will typically also carry microphones, and sensors for chemical, biological, or nuclear agents, so that the operator is able to decide what the 'bomb' is and how best to deal with it. Many such robots have a gripper on the end of the robot arm in case, for instance, a door needs to be opened, or to handle or move the suspicious device.

Unmanned air vehicles (UAVs)

Tele-operated aircraft are generally referred to as unmanned air vehicles, or UAVs. They may be fixed wing or rotary wing (helicopters), and in the military domain the best known (but controversial) UAVs are known as drones. In contrast with most other tele-operated robots, modern UAVs often have a high degree of autonomy. Fitted with a GPS receiver for satellite navigation and an autopilot, a UAV will typically be able to fly a route to a given destination, via a set of way-points, with little or no intervention from the remote pilot, much like a modern piloted commercial aircraft. UAVs will also typically be able to take off and land autonomously—something that is especially difficult to do remotely for a small aircraft. Thus a UAV operator is able to focus primarily on monitoring the vehicle's status and position, and, of course, on collecting pictures and surveillance data.

Planetary rovers

With a distinguished history stretching back to the Lunokhod moon rover of the 1970s, tele-operated robots for surface exploration—generally referred to as planetary rovers—have provided not only remarkable science from the surface of the moon and more recently Mars, but inspirational examples of robotics engineering. Planetary rovers are tele-operated mobile robots that present the designer and operator with a number of very difficult challenges. One challenge is power: a planetary rover needs to be energetically self-sufficient for the lifetime of its mission, and must either be launched with a power source or—as in the case of the Mars rovers—fitted with solar panels capable of recharging the rover's on-board batteries.

Another challenge is dependability. Any mechanical fault is likely to mean the end of the rover's mission, so it needs to be designed and built to exceptional standards of reliability and fail-safety, so that if parts of the rover should fail, the robot can still operate, albeit with reduced functionality. Extremes of temperature are also a problem, in particular the cold, since electronic components will not operate below about −50C. Thus rovers, like spacecraft, need to have heaters and insulation for their electronics.

6. Mars exploration rover

But the greatest challenge is communication. With a round-trip signal delay time of twenty minutes to Mars and back, tele-operating the rover in real time is impossible. If the rover is moving and its human operator in the command centre on Earth reacts to an obstacle, it's likely to be already too late; the robot will have hit the obstacle by the time the command signal to turn reaches the rover. An obvious answer to this problem would seem to be to give the rover a degree of autonomy so that it could, for instance, plan a path to a rock or feature of interest—while avoiding obstacles—then, when it arrives at the point of interest, call home and wait. Although path-planning algorithms capable of this level of autonomy have been well developed, the risk of a failure of the algorithm (and hence perhaps the whole mission) is deemed so high that in practice the rovers are manually tele-operated, at very low speed, with each manual manoeuvre carefully planned. When one also takes into account the fact that the Mars rovers are contactable only for a three-hour window per Martian day, a traverse of 100 metres will typically take up one day of operation at an average speed of 30 metres per hour.

Surgical robots

In complete contrast to planetary rovers, the operational domain of surgical robots is inner space. Should you be unlucky enough to need prostate surgery, there's a fair chance that you will be operated on by a robot or, to be precise, a surgeon tele-operating a robot. The benefits of minimally invasive, or keyhole, surgery are well known, but the precision and dexterity with which the surgeon can manually control the instruments are limited. A surgical robot overcomes this limitation through the use of tele-manipulation.

Typically a number of laparoscopic ports are opened through the abdomen, to the site that requires surgery. Then cameras and lighting are introduced via one of the ports so that the surgeon has a close-up and possibly 3D view of the workspace. Instruments, such as miniature scalpels, forceps, or suturing needles are introduced via the other ports so that the surgeon can undertake the procedure while looking through the cameras.

The instruments are tele-operated by the surgeon via a special pair of hand controls. Importantly, the robot's control system then scales down the surgeon's hand and finger movements so that a movement of perhaps 0.5 cm by the surgeon causes a movement of 0.5 mm in the surgical instrument. The robot may also filter out any trembling in the surgeon's hands so that the microsurgical instruments' movements are smooth and precise. Generally the surgeon's console is physically inside the operating theatre, along with the patient undergoing surgery and the whole surgical team; but, of course, it need not be: one of the benefits of this approach is that a surgeon may operate from a different location.

Robots for education

A significant milestone in the development of robots for education was the LEGO Mindstorms system, now in its second generation as LEGO NXT. Developed in collaboration with the MIT media lab, the LEGO NXT system is widely used to support the teaching

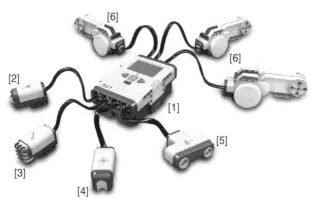

7. LEGO Mindstorms NXT 'central nervous system'

of robotics and is well regarded in the robotics community. At the heart of the system is a programmable control unit called the NXT intelligent brick (see Figure 7).

As shown here, the basic Lego NXT system has, in addition to the brick [1], three motor units and four different sensor modules. The motor units [6] are based on servomotors with reduction gearboxes and shaft encoders. The shaft encoders are able to sense the position of the output shaft to within 1 degree of accuracy, which means that they can be used as actuators to position the joint of a robot arm, for example, with good precision.

Sensor module [2] is a simple binary touch sensor, which outputs the value 0 if not pressed, or 1 if pressed. Sensor module [3] is an analogue sound sensor with an output between 0 for silence and 100 for very loud sounds. Module [4] is an analogue light sensor with an output between 0 and 100 (complete darkness to bright light). The module also incorporates an LED to light an object. The light sensor can be used on its own or in conjunction with the LED, to measure the brightness of reflected LED light and hence the distance to an object.

Sensor module [5] is an ultrasonic distance sensor able to detect the distance to an object up to a range of 233 cm, with a precision of 3 cm; the module can also sense movement. Additional sensors not shown here are available, such as a colour sensor and a temperature sensor, in addition to third-party modules including a compass and accelerometer.

Compare the image in Figure 7 with Figure 2 in Chapter 1, and what we see, as if laid out on a dissection table, is the complete 'central nervous system' for a LEGO NXT robot. We see the 'brain' and energy supply (in the brick), the 'sensors' and 'muscles' to provide mobility, and the wiring (i.e. the 'nerves'). What is missing here is the body—the structural components—and, of course, the software that would make a particular robot with a particular function.

Very simple programs can be entered directly using a simple menu system and buttons on the brick, but much more sophisticated programs can be created on a laptop or PC and downloaded to the brick. The range of programming tools (from beginner to expert) and the fact that LEGO has released an open-source licence for NXT hardware and software mean that this is a modular robotics development system of great value, not only in education, but also as a very flexible platform for testing ideas in research. An Internet search quickly reveals an extraordinary range of robot models that have been developed with LEGO NXT, including robot Rubik's cube solvers, a robot that balances on two wheels, and a fully automated LEGO NXT model factory that assembles LEGO cars.

A robot taxonomy

Getting a sense of a where any particular robot fits into the family tree of all robots is not easy. Some robots, in particular those in well established applications, have unambiguous descriptors like UAV (unmanned air vehicle, aka drone), but many robots tend to be labelled in a rather ad hoc fashion. The adjective *humanoid*, for instance, tells us a robot might have some human-like shape but, apart from that, the word is rather unhelpful. It doesn't tell us whether the robot is a walking humanoid or an immobile upper body, nor does it suggest what the robot might be used for, or whether it is tele-operated or autonomous.

But in robotics there are a number of generally accepted terms for classifying robots. Taken together, these form a loose taxonomy that is helpful for understanding how robots are related (not literally, of course), and where any given robot fits into the whole of robotkind.

These terms fall into six categories. First, *mobility*: fixed robots are immobile (although they may well, of course, exhibit movement, for instance the movements of a paint-spraying robot). Mobile robots are capable of self-actuated locomotion, with legs, wheels, wings, or whatever gives the robot mobility in its working environment.

The second major category, *how operated*, tells us whether the robot is tele-operated or autonomous. But unlike the previous category, the distinction is blurred; the label 'autonomous' covers a wide range of robots and different degrees of autonomy. The third category classifies a robot by its shape: it may be an anthropomorph, i.e. shaped somewhat like a human; a zoomorph, with an animal-like body shape; or what I call mechanoid, a robot that looks nothing like any human or animal.

The fourth category I propose is determined by *human–robot interactivity*. In other words, is the robot designed to interact with humans or not? This category is useful because, for human-interactive robots many other characteristics, especially safety and trustworthiness, become important. The fifth category, *learning*, tells us whether a robot has a fixed repertoire of behaviours or is able to learn and hence adapt its behaviour during operation.

The sixth and final category, *application*, I have included because it has become the convention in robotics to classify robots as either industrial robots or service robots. The Industrial Federation of Robotics (IFR) defines an industrial robot as 'an automatically controlled, reprogrammable, multipurpose, manipulator…for use in industrial automation applications', and (tentatively) a service robot as 'a robot which operates autonomously or semi-autonomously to perform services useful to the well-being of humans and equipment, excluding manufacturing'.

Table 2 shows these top-level categories and their divisions.

The range of applications in which we see service robots is very wide indeed. Some well established applications are listed

Table 2. A robot taxonomy

Category	Major types	Subtypes (if any)
Mobility	Fixed (immobile)	
	Mobile	Wheeled/tracked/legged/flying/swimming
How operated	Tele-operated	
	Autonomous	
Shape/Morphology	Anthropomorphic	Humanoid/Android
	Zoomorphic	
	Mechanoid	
Human–robot interactivity	Human-interactive	Or animal-interactive
	Non human-interactive	
Learning	Fixed behaviour	
	Adaptive	
Application	Industrial robot	Multi-axis manipulator for industrial automation
	Service robot	See Table 3

Table 3. Service Robot Applications and their classification

Service robot application	Example	Classification
Agriculture	Fullwood Merlin robotic milking machine	Fixed; autonomous; mechanoid; animal-interactive; fixed behaviour
Arts & entertainment	Engineered Arts Ltd Robothespian	Fixed; autonomous; humanoid; human-interactive
Assisted living	Exact Dynamics iARM (intelligent assistive robotic manipulator)	Fixed; tele-operated; mechanoid; human-interactive; fixed behaviour
Portering	Aethon TUG Automated Robotic Delivery System	Mobile; autonomous; mechanoid; human interactive; fixed behaviour

(Continued)

Table 3. Continued

Service robot application	Example	Classification
Cleaning	iRobot Roomba robot vacuum cleaner	Mobile; autonomous; mechanoid; non human-interactive; fixed behaviour
Driverless cars	None yet commercially available	Mobile; autonomous; mechanoid; human-interactive; adaptive
Hobby & education	Lego Mindstorms NXT	Classification depends on design
Exploration	NASA Mars Exploration Rovers Spirit & Opportunity	Mobile; tele-operated, mechanoid; non human-interactive; fixed behaviour
Prosthetics	Touch Bionics i-limb prosthetic hand	Fixed; tele-operated (via skin electrodes); humanoid; human-interactive
Research & education	Aldebaran Robotics NAO	Mobile; autonomous; humanoid; human-interactive; adaptive
Search & inspection (defence and disaster response)	iRobot PackBot	Mobile; tele-operated; mechanoid; non human interactive; fixed behaviour
Surgery	Intuitive Surgical Inc. da Vinci robot	Fixed; tele-operated; mechanoid; human-interactive; fixed behaviour
Tele-presence	VGo Communications Inc. VGo	Mobile; tele-operated; mechanoid; human-interactive
Therapeutic & companion robots	Intelligent Systems Co. Paro robot baby seal	Immobile; autonomous; zoomorphic; human-interactive; adaptive
Toys & leisure	Innvo Labs Pleo animatronic pet dinosaur	Mobile; autonomous; zoomorphic; human-interactive; adaptive

Robotics

in Table 3 together with representative examples of robots in each application and their classification within the taxonomy above.

This chapter has described a number of robots, and types of robots, that exist in the real world, from factories to school classrooms. It is estimated that there are currently more than eight million robots currently in use. The range of real-world robots and the jobs they do is extremely broad, and illustrates that there are few areas of human endeavour that have not, to some extent, already benefited from robotics technology. In some areas the impact has been very significant, such as manufacture, but in many others the potential of robotics is yet to be realized.

Chapter 3
Biological robotics

Since the late 1980s, there has been a profound change in the way we approach the design of intelligent robots, which is often characterized as 'biological inspiration', or just bio-inspired robotics. But this change goes much deeper than just inspiration; we only need to think of Leonardo da Vinci to realize that designers and engineers have been inspired by nature for centuries. In intelligent robotics, bio-inspired means two things.

First, it means root-and-branch revision of the way we approach robot design, no longer treating it as an exercise in pure engineering but instead as a multidisciplinary endeavour overlapping with the life sciences. Intelligent robotics research very often now encompasses biochemistry, biology, neuroscience, and animal ethology.

Second, robots are not just inspired by nature but are instead directly modelled on, or in some cases are models of, designs or systems found in nature. The latter are known as biomimetic robots. This means that we should no longer think of robots as simply machines to do a job, but—in some very limited sense—models of living systems. Artificial animals, perhaps. In fact, a new discipline has grown up that is known as Artificial Life, or just *Alife*, that makes use of either computer simulation or real, physical robots to model living systems or processes.

How to build an autonomous robot

Take two very simple low-power motors, with wheels, and mount them on either side of a small robot chassis. Then take two low-cost solar panels and place one solar panel on each side of the robot so that they look very much like wings. Then connect (wire) the output of the left-hand solar panel to the right-hand motor, and vice versa. Robotics engineer Chris Bytheway built such a robot.

Called Solarbot, this little robot behaves in the following way. A strong light that shines with equal intensity on both of the robot's wings will power both motors, more or less equally, and the robot will drive forward, toward the light source. If, on the other hand, the right-hand wing is in darkness but the left-hand wing is illuminated, then only the right-hand motor will have power; the left-hand motor will be stopped and the robot will therefore turn to its left—as if it were trying to get into the light. The opposite situation—left wing in darkness and right wing illuminated—will cause the robot to turn to its right. As a result, the robot is able to find its way through a simple obstacle course, toward the light. Solarbot is said to demonstrate positive phototaxis and a simple kind of obstacle avoidance. Solarbot also orients itself toward the light if it's not directly facing it.

Solarbot provides us with a remarkable illustration that simple, apparently purposeful behaviours require no computational machinery at all. Solarbot is an example of what roboticists refer to as a Braitenberg machine.

In his 1984 book, *Vehicles*, Braitenberg showed, with a wonderful set of thought experiments, that just cross-connecting sensors and actuators in the way I have outlined here could achieve simple reflexive behaviours of surprising sophistication. What Solarbot also illustrates is that a simple robot without (in a sense) a brain can show interesting and useful behaviour while at the same time illustrating that the behaviours are a property of the interaction between a robot and its working environment. The robot's behaviours are said to be an *emergent* property of those interactions.

Furthermore, the complexity of those behaviours is linked to the complexity of the environment. So Solarbot in a dark room does nothing at all. In an empty room with a single light source but no obstacles, the robot's behaviour is marginally more interesting: it will simply go toward the light source until it crashes into it (fruitlessly then continuing to 'feed' on the solar energy and drive its motors forward forever). Only with a set of obstacles that create patches of shadow do we see Solarbot's full repertoire of behaviours.

Solarbot also shows us that an *autonomous* robot doesn't have to be complicated. In fact, counter-intuitively, autonomous robots are often simpler than tele-operated robots (imagine how much more complicated Solarbot's control systems and electronics would need to be for you to be able to steer it through the maze by radio control). The realization that the behaviour of an autonomous robot is an emergent property of its interactions with the world has important and far-reaching consequences for the way we design autonomous robots.

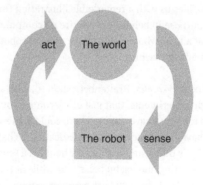

8. The robot–world feedback loop

Figure 8 illustrates what roboticists call the outer control loop, for autonomous robots. The loop is said to be: 'closed through the robot's working environment'. What this means is that the robot senses its environment, and the values received by its sensors in some way result in new motor outputs (directly in Solarbot's case). The robot then moves (all or part of itself), and—in all likelihood—as a result of those moves, the robot's sensors will receive different values. Those new sensor values will perhaps generate different behaviours, which in turn cause the robot to move, and so on. In this way the sense–act loop of the robot is closed. What does this mean for roboticists?

Well, first, when we design robots, and especially when we come to decide what behaviours to programme the robot's AI with, we cannot think about the robot on its own. We must take into account every detail of the robot's working environment. If that environment is simple, in the sense that the designer can measure or specify more or less everything that the robot will encounter and interact with (and how those objects themselves will behave), then the job of designing the robot and its behaviours may not be too challenging. If, on the other hand, the environment in which the robot has to operate is unknown (like the surface of an unexplored planet), or known but unpredictable, then it becomes not just difficult but impossible to predict every possible situation the robot might find itself in, and therefore design for all of those eventualities. The latter case, *known but unpredictable*, covers, of course, pretty much every human environment in which we might need robots: human living, play, and work spaces, including roads and outdoor urban spaces.

Second, it makes it difficult to test an autonomous robot, and especially to prove that it will always behave as expected (remember, those behaviours are emergent properties of robot-working environment interaction). When you place even a simple robot with completely determined behaviours (like Solarbot) in an unpredictable environment, then it becomes impossible to predict exactly what the robot will do. This is not a problem for lab robots like Solarbot, but if the robot is required to work, for instance with humans, then proving (and certifying) that the robot will always be safe becomes a very significant challenge.

Autonomous robots in unpredictable (especially human) working environments are, of course, still a new and emerging technology, but it may be that new approaches to design and certification are needed — approaches that perhaps depart radically from traditional engineering practice in which everything is precisely specified, designed, and tested. The natural world is a complex place and most simple animals make a living without being 100 per cent aware of everything that's going on around them, or having behavioural responses for every possible eventuality. A bio-inspired approach would suggest that autonomous robots are designed to be good enough, and safe enough, to do the job required of them, but no more.

Behaviour-based robotics

In the 1970s, experimental autonomous mobile robots moved slowly and hesitatingly through the obstacles set up for them to navigate. Their slowness was nothing to do with weak or ineffective motors or locomotion. It was because these robots, exemplified by Stanford Research Institute's remarkable robot Shakey, had control systems based on the then prevailing approach to artificial intelligence, known as symbolic representation.

Robots such as Shakey used a *deliberative* control approach known as Sense–Plan–Act. The idea was that the robot would use data from its sensors to build and maintain an internal representation of its world. The robot would, for instance, capture an image with its camera and, using image processing, identify the objects in its field of view, work out where the corners and edges are, and the gaps between different objects — hence building a kind of spatial map of what it 'sees'. The robot would use this internal map to plan its next move before then actually executing this move, to complete one Sense–Plan–Act (SPA) cycle. Then, after the robot has moved forward by perhaps a few centimetres, it would again capture an image and start the SPA cycle again.

Herein lies the problem. Because the robot has moved, its view of the world is now different, and so it must do some rather complex calculations to deduce, for instance, that the red block on the left is the very same red block it 'saw' in the previous image (for its view of that object has changed), and hence both maintain its internal map and update its own position and orientation in that map.

It's easy to see that this is a complex and slow process, even for completely static environments where the only thing moving is the robot itself. A moving object in its field of view, such as a person, could easily render the robot helplessly immobile, unable to reconcile what it 'sees' with its internal map and plan its next move. Contrast this with Solarbot, that does no planning at all: its control system is just Sense–Act, or *reactive*.

In the mid 1980s, MIT roboticist Rodney Brooks realized that the prevailing approach to autonomous mobile robot control must be wrong. His thinking went along the following lines. Most animals on the planet, insects for instance, cannot possibly have the cognitive machinery to build and maintain internal representations of the complex, dynamic environments in which they have to make a living. Yet these animals are highly successful at navigating their environment, foraging for food, finding mates, and avoiding predators. They don't need an internal representation because they have the world itself. The world is, as Brooks wrote 'its own best model'. Thus a mobile robot, just like an insect, should be able to move around in the world, avoiding collisions with static obstacles while safely getting out of the way of moving objects, simply by sensing those objects and reacting only when they get close enough to pose a threat.

In one of the most influential papers of modern robotics, 'A robust layered control system for a mobile robot', Brooks proposed an alternative to the Sense–Plan–Act approach, in which a robot's control system is built, from the bottom up, with a set of discrete task-achieving 'behaviours'. Each of these layers accepts input from the robot's sensors and is capable of driving the robot's motors. Upper layers represent higher levels of competence and all of the layers work in parallel.

Depending on what overall work the robot is designed to do, high-level layers might well be required to build maps of the robot's working environment; but crucially, low-level behaviours such as obstacle avoidance will not rely on those maps. Thus a robot designed along these lines will be able to move around safely, reacting quickly to obstacles or hazards, yet at the same time achieving higher-level goals such as getting from one place to another.

The behaviour-based approach combines the strengths of the reactive and deliberative approaches, which are appropriate to different layers of the robot's controller. Importantly, the behaviours are independent. They do not, for instance, all need to access a world map. A high-level behaviour may well build and maintain a map; a low-level behaviour doesn't need that map: instead, it will react directly to sensor inputs. Because of the independence of the behavioural layers, the overall observed robot behaviour—and its functionality—results from *both* the interaction of the robot's individual behaviours *and* the interaction of the robot and its environment.

Maja Matarić's neurobiologically inspired robot Toto provides a very nice example of this approach. A low-level behaviour called 'Boundary tracing' provided Toto with collision-free wandering while finding boundaries (i.e. walls) and following them. A mid-level behaviour called 'Landmark detection' tracked Toto's sensor readings and recognized landmarks such as a wall on the robot's left, or on its right, or on both sides, i.e. a corridor. The high-level behaviour 'Goal-oriented navigation and map learning' learned this sequence of landmarks and hence built a map, allowing Toto to find its position in the map and navigate to new positions. Rats navigate by following boundaries and Matarić argued that Toto's AI might help in understanding the part of the rat's brain responsible for navigation: its hippocampus.

Slugbot: a robot predator

Like all machines, robots need power. For fixed robots, like the robot arms used for manufacture, power isn't a problem because the robot is connected to the electrical mains supply. But for mobile robots power is a huge problem because mobile robots need to carry their energy supply around with them, with problems of both the size and weight of the batteries and, more seriously, how to recharge those batteries when they run out.

For autonomous robots, the problem is acute because a robot cannot be said to be truly autonomous unless it has energy autonomy as well as computational autonomy; there seems little point in building a smart robot that 'dies' when its battery runs out.

For outdoor mobile robots, one obvious approach is to use solar panels to directly charge a robot's batteries. Sunlight then becomes the robot's energy source. The Mars rovers Spirit and Opportunity are a dramatic example of the success of this approach and these robots could truly be described as energetically autonomous.

However, relying on sunlight alone could be a problem in winter when nights are long (Spirit and Opportunity have to in effect hibernate during the Martian winters), especially if we want the robot to operate day and night and in all weathers. Would it be possible for outdoor robots (on Earth, not on Mars) to get energy by actively foraging or predating, like animals? In 1998, in the Bristol Robotics Lab, a research project to investigate this possibility began, and its first robot was called Slugbot.

Slugbot was the first attempt to build a robot predator: a robot able to catch slugs and then use those slugs for energy. Slugs were chosen as the food source for two reasons: they are slow-moving and therefore (in theory) easy to catch, and they are regarded by almost everyone as pests. Designed by Ian Kelly, Owen Holland, and Chris Melhuish, Slugbot searches for and catches its prey with a long, lightweight carbon-fibre arm. At the end of the arm is a special-purpose camera/gripper for both sensing and collecting the slugs.

This ingenious arrangement allows the robot to thoroughly search the ground around it for slugs by scanning in a circular, spiral pattern. The gripper, with the camera in its 'palm', is held a few centimetres above the ground while scanning, which means that when a slug is found, the arm only has to lower the gripper onto the soil directly below it to grab the slug. A robot predator would have to be extremely parsimonious about using up the energy gleaned from its food. This is reflected in Slugbot's design: moving its carbon-fibre arm costs much less energy than moving its body, so relative to its size Slugbot can search a large area of ground for its prey, using minimal energy.

Slugs are hard to detect; they are the same temperature as the soil, so we cannot use a heat sensor to find them. Nor can we simply use a camera to look for something slug-shaped, otherwise stones, leaves, or pieces of wood that happen to be slug-shaped would be falsely identified as slugs. Slugbot's designers solved this problem by noticing that slugs reflect red light. They placed a red light source, together with the camera, in the palm of Slugbot's gripper. The robot's image-processing software then looks for red slug-sized and -shaped blobs.

Slugbot solved the problem of predation (for slugs at least) but was unable to convert the slugs into electrical energy—a problem addressed by Slugbot's successor: EcoBot.

EcoBot: a robot with an artificial digestive system

There are several different approaches to converting biomass into electrical energy. One is to ferment the biomass, generate biogas (methane), then burn the methane in a fuel cell. Another conceptually much simpler approach is to make use of a microbial fuel cell. A microbial fuel cell, or MFC, is rather like a battery except that instead of generating electrical energy from a chemical reaction, it makes use of a biochemical reaction.

Just like an ordinary battery, an MFC has two sections: the anode and the cathode. But in the MFCs developed in the Bristol Robotics Lab (BRL), the anode compartment contains a liquid 'soup' of microbes capable of literally digesting food. Typically this might consist of sewage sludge, chosen because it contains a broad spectrum of bacterial species capable of digesting (and breaking down) more or less any biological material.

A side effect of that process of digestion is that ions are produced which, of course, have an electrical charge. The cathode in the BRL MFCs is open to air, and oxygen is absorbed, which helps to exchange electrons across a membrane between the anode and cathode. If the MFC is connected within an electrical circuit, then a current flows from the anode to the cathode, and we have a working biological battery that runs on food.

MFCs generate very low levels of power—typically a single MFC provides just a few microwatts (1,000 times less than a standard AAA cell), and during the early development of the BRL MFCs, researcher Ioannis Ieropoulos tested different foodstuffs to try and find one with the best energy efficiency. The material he discovered worked best is the polysaccharide chitin, found typically in insect exoskeletons. Remarkably, an early prototype of the EcoBot was powered by only eight dead houseflies (Musca domestica), one per MFC.

Eight MFCs wired in series generate too little power to continuously operate a robot, so the electrical energy generated by the MFCs is used to charge a bank of super-capacitors until there is enough saved energy to power the robot for action. EcoBot thus operates in a pulsed mode in which it cycles between periods of inactivity, while it is digesting food, and rather shorter bursts of activity when it senses its environment, transmits data, and drives its motors. Thus EcoBot II was not only the world's first dead-fly-powered robot, but also perhaps the slowest, moving at about 13 cm per hour. However, fuelled by just eight dead flies, it operated continuously for nearly two weeks.

While it was a considerable achievement, EcoBot II had two serious drawbacks. The obvious one was that it had to be 'fed' by hand; in other words, the food was manually introduced into the anode chambers of the eight MFCs. Unlike its ancestor Slugbot, EcoBot II was unable to capture its own food. A second, equally serious, problem stemmed from the design of the MFCs and in particular the closed anode chamber. Digestion produces waste products and, eventually, those products will poison the digester micro-organisms, so the MFC will stop working until its digester inoculum is refreshed. EcoBot III was designed by Ioannis Ieropoulos, John Greenman, Chris Melhuish, and Ian Horsfield to overcome both of these limitations.

A critical innovation in EcoBot III is in the design of the individual MFC. As shown in Figure 9 (inset), input and output ports have been added to the anode chamber so that the MFC's microbial inoculum can be cycled. Also important is that the MFC has been shrunk to around 1 cubic cm; tests showed that a larger number of smaller MFCs is the best way to increase the power output of the robot's artificial digestion system. EcoBot III is fitted

9. EcoBot III: a robot with an artificial digestive system. The inset image shows a single MFC

with a total of forty-eight MFCs in two rings around the body of the robot.

In another innovation, EcoBot III has a flytrap at the top of the robot. Inspired by the pitcher plant, a combination of colour and artificial pheromones attract flies to enter the trap, where they fall into a pond containing the microbial inoculum. In the pond, the flies start to be digested and result in a nutrient-rich solution, which is then transferred via tubes to the forty-eight MFCs, where the digestion process is completed and energy generated to power the robot. To extract the maximum amount of energy, the inoculum is recycled twice through the MFCs. Solid waste products are filtered from the solution and, literally, excreted by the robot. EcoBot III thus represents the only known example, to date, of a robot with a complete artificial digestive system.

The robot's experimental test environment consists of a closed glass tank into which live flies are introduced. The robot runs on wheels and moves back and forth along a short set of rails, from one end of the tank to the other. This serves to demonstrate that the robot is capable of generating enough energy for locomotion, over and above that required to operate its own internal metabolism. In fact, for energy autonomy, the robot requires hydration as well as protein and is able to collect water from a water tube at one end of its track. Providing there is a continuous supply of water and protein in the tank (in the form of flies), then, theoretically, the robot should be capable of continuous operation (unlike its predecessor EcoBots).

EcoBot III is, I believe, significant for several reasons. First, it provides a proof-of-concept demonstration of energetically autonomous robots that collect and digest food for energy, and can do so while generating enough surplus power to do work. Second, EcoBot III demonstrates the novel application of design and fabrication approaches, making extensive use of 3D printing for complex plastic shapes and structures (that integrate some of the robot's plumbing, for instance). Third, the robot demonstrates a cross-disciplinary collaboration between robotics and biochemistry. While it might be regarded as only rather loosely bio-inspired (by the pitcher plant), EcoBot III can truly be described as a biological robot.

What are the potential applications of EcoBot technology? One could imagine, for instance, a gardener robot that is able to identify and selectively pick weeds and larger pests (like slugs, in fact), power itself with what it has collected, and then excrete its waste as fertilizer. Such a robot would in effect 'live' in your garden, quietly and autonomously getting on with its work. In horticulture and agriculture one could imagine groups of such robots self-organizing (see swarm robotics in Chapter 5) to collectively—and organically (since no herbicide or pesticide is involved)—control weeds and pests in fields and greenhouses.

Scratchbot: a biomimetic approach to sensing the world

Many animals use active touch to sense their environment. Insect antennae and mammalian whiskers are both examples of these so-called active touch sensory systems, or mechano-sensors. Rodent whiskers are particularly interesting since their whiskers actively 'whisk' back and forth, gathering information about the position, shape, and texture of objects within the whiskers' radius.

Rats, for instance, see in the dark in 3D with texture, using their whiskers, and neuroscientists have estimated that the rat uses a greater proportion of its cortex for processing data from its whiskers than from its vision system. There seems little doubt that the predominant sense for the rat is active touch using its whiskers.

Artificial whiskers are very attractive sensors for robots. Imagine small search-and-rescue robots designed to enter collapsed buildings to search the spaces inside the structure. Those spaces are likely to be dark, and quite possibly the air will be laden with dust or smoke particles. Thus conventional robot sensors based on light, such as camera-based vision systems or laser-based light detection and ranging (LIDAR), will be more or less useless.

Artificial whiskers might, in principle, provide the best form of sensor to enable the robot to navigate and search such confined and dangerous spaces. In such an environment a robot would have to tread carefully, and the feather touch of artificial whiskers would minimize the risk of the robot itself causing any further collapse.

The Bristol Robotics Lab Scratchbot (see Figure 10) was designed and built as a testbed for artificial whiskers. Designed by Martin Pearson, Tony Pipe, and Jason Welsby, within the BIOTACT project led by cognitive neuroscientist Tony Prescott, Scratchbot is

10. Scratchbot, a robot with artificial whiskers

about twice the size of a rat, and its 'head' has two sets of large whiskers, arrayed on either side. These whiskers, which model the large whiskers (macro-vibrissae) of the rat, are arranged as three rows of three whiskers. Of course, the real rat has far more than eighteen such whiskers. Motors allow all eighteen of the robot's macro-vibrissae to whisk back and forth, providing a sensing radius of approximately 20 cm and an arc of about 60 degrees on either side of the robot's head.

Each individual artificial whisker is about 20 cm in length, with a circular cross section tapering from about 2 mm diameter at its base to 0.6 mm diameter at the tip; it is 3D-printed from a plastic material that gives the whisker strength, lightness, and flexibility. The base of the artificial whisker is firmly set into a root, into which is integrated a two-axis Hall-effect sensor. This sensor, analogous to the mechano-receptors in the follicle of a real whisker, generates electrical signals proportional to the amount and direction of flex of the whisker.

At the front of the robot's head is another set of artificial whiskers arranged in a 4 × 4 grid pattern. These shorter whiskers are fixed (i.e. they do not actively whisk) and they model the rat's small whiskers (micro-vibrissae). It is believed that the rat uses its macro-vibrissae to first locate objects within its 'touch' field of view, then turns its head to position the micro-vibrissae on the object in order to examine the object with greater precision; the micro-vibrissae thus act as a touch sense equivalent of the foveal region of the eye. Scratchbot emulates this behaviour. It will search a region using its whisking artificial macro-vibrissae and when the robot finds an object it will then, using its articulated neck, turn its head to 'focus' its micro-vibrissae on the object.

How does Scratchbot make sense of the signals from its artificial whiskers? Even though Scratchbot has far fewer whiskers than its biological counterpart, with thirty-four whiskers there is a significant amount of data dynamically generated by an encounter with an object. Transforming that data into the object's position and orientation (relative to Scratchbot's body) or its shape and texture is not at all straightforward.

An artificial whisker generates no signal at all when it is static. Data is generated when the whisker contacts the object, and the nature of the data generated as the whisker scrapes across the surface of the object can be used to infer its texture: a rough texture will, for instance, generate a different signal from a smooth texture (in the same way that you cannot tell if a surface is rough or smooth simply by placing your finger onto the surface— you actually need to draw your finger across the surface).

When all of the whiskers in the macro-vibrissae array move across a single object, the relative timing of the signals generated by the whiskers, when correlated with the relative position of the whiskers generating those signals, can be used to infer the position of features, especially corners and edges, on the object in 3D space.

Scratchbot's designers took a radical biomimetic approach to the way in which the robot processes the data generated by its artificial whiskers. The robot has an electronic model of a very small part of the rat's brain: the part that preprocesses the nerve impulses from the whiskers. The robot implements, using field-programmable gate array (FPGA) technology, a high-fidelity spiking artificial neural network model of a part of the rat's brainstem, called the trigeminal sensory complex. The FPGA initially processes the signals from the whiskers using about 40,000 artificial neurons and the same connection structure.

Scratchbot therefore provides us with a good example of a strongly biomimetic approach to robot design. It illustrates the way in which robotics is now a cross-disciplinary endeavour—in this case between roboticists, animal ethologists, and neuroscientists. The robot is a working model to test scientific hypotheses—in this case about how certain control systems in the rat's brain work—while at the same time providing a testbed for the development of a new robot sensing technology. Artificial whiskers of the type demonstrated by Scratchbot are without doubt a powerful addition to the range of sensors available to robot designers.

Robots that learn

Learning is clearly a powerful asset to any creature, be it animal or robot, and roboticists have, not surprisingly, investigated approaches to robot learning for many years. But few real-world robots are able to learn or adapt their behaviours. This is a reflection of two things: first, that robot learning has been and remains a significant research challenge; and second, that robots able to learn pose a particular problem when it comes to certifying the safety of a robot. How can we certify a robot will always do the correct thing when we don't know what its future learned behaviours might be?

Robot learning is, however, important and here I summarize the main approaches.

The first, and without doubt most powerful and successful, robot learning approach (which some regard as one of the greatest achievements in robot science to date) is called 'Simultaneous localization and mapping', or SLAM. Localization is a major problem in mobile robotics; in other words, how does a robot know where it is, in 2D or 3D space?

If it's an outdoor ground or flying robot, then the global positioning system, GPS, provides a ready solution. If it's an indoor robot, such as the Aethon TUG hospital portering robot I described in Chapter 2, and the robot is provided with an electronic map, then localization is relatively straightforward.

However, for robots that have neither GPS (or something like it) nor a map, SLAM provides a technique for building a map (of what the robot can see with its sensors) while approximating its own position in that map at the same time. As the robot moves though its environment, the map and the robot's level of confidence of its position relative to the objects in that map improve.

A more general type of robot learning is called reinforcement learning. This is certainly bio-inspired, since it is a kind of conditioned learning. If a robot is able to try out several different behaviours, test the success or failure of each behaviour, then 'reinforce' the successful behaviours, it is said to have reinforcement learning. Although this sounds straightforward in principle, it is not. It assumes, first, that a robot has at least one successful behaviour in its list of behaviours to try out, and second, that it can test the benefit of each behaviour—in other words, that the behaviour has an immediate measurable reward. If a robot has to try every possible behaviour or if the rewards are delayed, then this kind of so-called 'unsupervised' individual robot learning is very slow.

Also bio-inspired, particularly by social learning in humans, is a third approach to robot learning, called robot *programming by demonstration*. In Chapter 2 I explained how a user may program a robot arm to go through a required sequence of movements with a control pad that allows the human to control the robot arm and—in effect—lead it 'by the hand' and thus 'teach' the robot. This represents a very simple form of programming by demonstration, but in recent years the focus has shifted to learning by imitation.

Learning by imitation is especially appropriate to humanoid robots and, in essence, it involves a human undertaking a task while the robot watches. The robot then attempts to copy the same movements and actions. Programming this form of imitation learning is complex and requires the solution of the so-called correspondence problem: the robot has to transform what it 'sees' with its cameras into a corresponding sequence of motor actions.

The three approaches I have outlined above are all methods for an individual robot learning a task or skill. In nature, most simple animals do not learn individually, but are born with a genetically determined set of reflexive actions. However, those behaviours have adapted over successive generations, through natural selection. Inspired by Darwinian evolution, we have, in robotics, a fourth type of learning, sometimes called evolutionary learning—which takes place over multiple generations of robot. I describe this approach in Chapter 5.

Biological robotics is not some Frankenstein fantasy to create 'living machines'. In contrast, the profound change I am describing here has brought with it a deeper sense of humility and respect in the face of the exquisite complexity of the evolved 'designs' we see in nature. Our best efforts to build artificial models of these designs are pitifully limited, and serve to illustrate the gulf between the most complex intelligent robot and the simplest animals.

Chapter 4
Humanoid and android robots

Although the vast majority of robots in the world today—including those in research labs—are not humanoid, robots made in our likeness hold a special fascination. This is perhaps not surprising given that the word 'robot' was first used to describe a fictional humanoid robot. But fiction aside, there are good reasons why robots that need to work with people might have to be humanoid: first, so they can use human tools and share human workspaces; and second, to be able to communicate naturally with humans (for instance, through speech, facial expressions, and gestures).

A robot is described as humanoid if it has a shape or structure that to some degree mimics the human form. Thus a robot head, with two vision sensors in approximately the correct place for eyes, positioned above a torso, would be regarded as humanoid. If the robot has arms, then these would similarly need to be humanoid, attached at the shoulders of the torso, with hands or grippers that approximate hands.

For a robot to be called humanoid, its form is more important than the detail of its components. If it has legs, for instance, it must be bipedal with some kind of hip joint, knee joint, and ankle joint—even if the legs may bear very little anatomical resemblance to their human counterparts. Most humanoid robots are decidedly mechanical in their appearance and if formed of plastic components they may appear more like cartoon people or even cartoon impressions of what robots should look like.

A small subset of humanoid robots does, however, attempt a greater degree of fidelity to the human form and appearance, and these are referred to as *android*. They have artificial skin and hair, make-up and clothes—mannequins, in fact. But like so many science fiction robots, peel away their artificial skin and underneath are the circuit boards and motors of the robot. Except that these robots are real, the closest yet to what is for many the holy grail of robotics: artificial people. Unfortunately (or perhaps fortunately, depending on your point of view), today's android robots—although superficially impressive—are far short of that dream.

In a nutshell the problem is this: we can build the bodies but not the brains. It is a recurring theme of this book that robot intelligence technology lags behind robot mechatronics—and nowhere is the mismatch between the two so starkly evident as it is in android robots. The problem is that if a robot looks convincingly human, then we (not unreasonably) expect it to behave like a human. For this reason whole-body android robots are, at the time of writing, disappointing.

Although humanoid robots are compelling subjects for research and development, they present particular technical challenges and it is not surprising that there are few in real-world use. Perhaps the first serious deployment of a humanoid robot as a robot assistant is the NASA GM Robonaut, recently installed on the International Space Station. This has a human-sized robot upper-body torso with arms and dexterous hands, and a head that looks rather like a racing driver's helmet.

The great majority of humanoid robots are found among toys (such as WowWee's Robosapiens), in education (such as the Aldebaran Robotics' Nao robot), or as technology demonstrators (such as Honda's ASIMO). Of particular interest is the growing use of humanoid robots in robot sports, where the robots might compete in single-competitor events (marathon running or weight lifting, for example), or in team events such as Robo Soccer. Robot football provides a remarkable challenge and showcase for robotics development, and small hobby humanoid robots are inexpensive enough that school or student teams can participate. Watching two teams of humanoid robots battling it out on a soccer pitch is a fascinating spectacle. But some of the most interesting humanoid robots are to be found in research labs, and I shall describe a number of them in this chapter.

It is important not to overstate the case for humanoid robots. Without doubt, many potential applications of robots in human work- or living spaces would be better served by non-humanoid robots. The humanoid robot to use human tools argument doesn't make sense if the job can be done autonomously. It would be absurd, for instance, to design a humanoid robot in order to operate a vacuum cleaner designed for humans. Similarly, if we want a driverless car, it doesn't make sense to build a humanoid robot that sits in the driver's seat. It seems that the case for humanoid robots is strongest when the robots are required to work alongside, learn from, and interact closely with humans.

Applications for humanoid robots: robot workplace assistants

One of the most compelling reasons why robots should be humanoid is for those applications in which the robot has to interact with humans, work in human workspaces, and use tools or devices designed for humans. Imagine, for instance, that you have a robot workmate. You are working to repair some complex machinery and those repairs demand your complete attention. But you need a tool, so you put your hand out and, without turning your head, ask 'Could I have the 15 mm spanner, please?' To be useful, your robot workmate needs to be able to understand both your spoken request and your gesture; in particular, the robot needs to know that you implicitly want it to put the 15 mm spanner in your hand—gently—and then let go of the spanner once you have grasped it. In noisy environments there is even more reliance on gestures and non-language utterances like 'Uh-huh!'

Of course, a robot that can understand your verbal and gestural language doesn't itself need to be humanoid, but being humanoid will help a robot in being able to physically locate and grasp the spanner from the toolbox or tool rack that was designed for humans. But there are stronger reasons than this. A robot that has a similar body plan to ours and is constrained to move in the same way as we move, and that sees or hears the world from a similar perspective to ours, is more likely to be able to learn both our ways and our workplaces.

BERT (see Figure 11) provides us with a case study of a robot designed for research in cooperative human–robot interaction that could eventually lead to the kind of humanoid workplace assistant robot imagined here. BERT's capability to interact with humans was developed at the Bristol Robotics Lab (BRL) as part of the CHRIS project by roboticists Chris Melhuish, Tony Pipe, Alex Lenz, and Sergey Skachek.

Designed by robotics company Elumotion, BERT II is a fixed robot but its lack of independent mobility simply reflects its function as a platform for research in close-proximity human–robot interaction. There is no reason in principle that BERT, or a descendent, couldn't walk on legs or roll on wheels. From its waist to the top of its head, BERT is about 1.1 m tall. Its torso is not actuated—in other words, it has a stiff, immobile back. BERT's arms are about 75 cm from shoulder to fingertip and have shoulder, elbow, and wrist joints which, although not anatomically accurate, provide BERT with a convincing emulation of a human's reach, range of arm movements, and — importantly — expressive arm gestures.

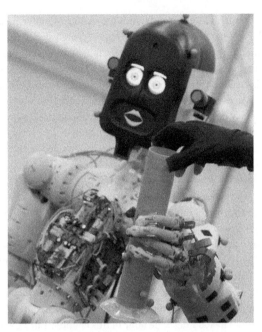

11. BERT II torso fitted with the BRL digital head, an upper-body humanoid robot for research in human–robot cooperation

In human gestural communication, hands are especially important; think of the special meaning of a tightly clenched fist, or the very different meaning of a clenched hand with a single outstretched pointing index finger, or an open hand palm-facing in greeting, or a very similar but insistent hand on a rigidly out-held arm signalling 'Stop'. Capable of all of these gestures, and more, BERT's hands are without doubt its tour de force. About the size of a large adult's hands, these artificial hands are a remarkable example of the humanoid robot builder's art. Electrically actuated, and with all of the motors embedded within the hand so that there is no need for any associated motors or artificial muscles in the forearm, or complex wrist joint with tendons, BERT's hands are capable of most natural hand movements.

Given that BERT's arms and hands are so anthropomorphic, it will come as no surprise that they have been designed for use as human prosthetic devices as well as for robotics. This illustrates the potential for very beneficial crossover between humanoid robotics technology and human prosthetics, and this is especially true for arms and hands. The requirements of the ability to emulate natural human movements and capabilities, low weight, low power requirements, reliability, and built-in low-level control (of individual joints and motors) have a good deal in common.

Let us now look at BERT's digital head. It is a plastic 3D structure with very approximate physical features. Built into the digital head is a flat LCD screen and its mouth, eyes, and eyebrows are drawn as computer graphics onto the flat screen. BERT's 'cartoon' face is thus realized with no mechanical moving parts at all, yet still allows for a wide range of static and moving facial expressions. Of course, BERT's head is incapable of the nuanced and high-fidelity whole-face expressions of the android heads I describe later in this chapter. But I believe that the approach illustrated in the BERT robot provides sufficient facial expressivity for a very wide range of potential real-world applications of humanoid robots that require artificial faces.

Humanoid robots need faces for two reasons. First, we humans instinctively focus our attention on the head and eyes of a robot (even when the eyes are not functioning), and a completely blank featureless face or no head at all can be unsettling. Second, the head and face can be a simple yet surprisingly powerful and natural way of providing feedback to the human interacting with the robot. If you verbally address a robot and it directs its eyes toward you, for instance (even if they are simple graphic eyes like BERT's), you have immediate confirmation that the robot has heard and is attending to you.

Shared attention through gaze provides another good example: if you and a robot are working on a shared task, you will naturally (and subconsciously) look at an object in the shared workspace. If the robot tracks your gaze, it can then focus its sensors on the same object and you and the robot have then established shared attention; if you see that the robot's eyes are looking at the same object, you then have confirmation.

Toward robot companions: robots that we can relate to

We humans have a deeply rooted propensity to anthropomorphize. We find it easy, in fact almost can't help, imbuing human characteristics to animals and describing almost anything that moves with language rooted in human emotions. Even a distinctly non-humanoid robot — like the e-puck robot introduced in Chapter 1 — will, for instance, be described as 'poorly' if it moves oddly compared with its teammates, or as 'aggressive' if the robot is constantly approaching others with fast, jerky movements. As a roboticist, I have learned to pay attention when observers use language like this, for it almost always means that those robots have developed a fault: a mechanical fault in a motor or gearbox, for instance, that gives the robot a kind of limp.

This tendency to anthropomorphize means that robotists do not, in fact, need to work very hard to build robots that seem to be behaving *as if* they have feelings. Robots don't have to be high-fidelity androids. Simple humanoid robot faces with cartoon-like features have been found to be perfectly adequate for allowing an emotional dialogue between human and robot. Of course, this dialogue is entirely one-sided: the human partner is doing all of the emotional work. Kismet, for instance, developed by Cynthia Brazeal at MIT, has large motorized eyes together with simply fashioned (and also animated) eyebrows and a mouth (see Figure 12).

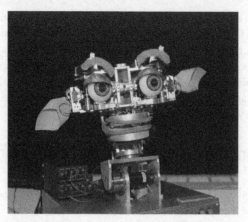

12. Kismet, an expressive robot head for research in social human–robot interaction

Kismet is a humanoid robot head and is particularly interesting because, despite appearing distinctly mechanoid, with all of its metal components and motors fully visible, it demonstrates how effective and compelling social interaction between robot and human can be.

Kismet's eyes are cameras that are able to move, and its neck, jaws, ears, eyebrows, and eyelids are all actuated. The cameras are interesting because Kismet is equipped with four: two wide-field cameras that the robot uses to decide what to pay attention to in its field of vision, and two narrow-field 'foveal' cameras, mounted in the robot's eyeballs, which focus that attention. The robot also has a microphone, which drives a speech-recognition system, and a speech-synthesis system driving a loudspeaker, so Kismet can recognize and respond to speech. Albeit somewhat minimalist, Kismet has all of the ingredients for engaging humans in natural and expressive face-to-face communication.

Kismet is programmed with a sophisticated AI that responds to human facial and verbal cues and provides the robot with a set of artificial 'emotions', including anger, disgust, fear, joy, sorrow, and surprise, designed to model infant–carer social interactions. As a result, Kismet responds to social interactions with a human in a way that invariably endears Kismet to them, and to anyone watching.

If we humans imbue relatively simple humanoid robots with feelings, consider how much more powerful the illusion if the robot is carefully crafted to mimic human facial expressions with a high degree of realism. David Hanson designs android robot heads that allow roboticists to experiment with artificial emotions, or, to be more accurate, to mimic the outward facial appearance of emotional states.

Designed and built by Hanson Robotics, Jules is an example of an android robot head created to mimic human facial expressions, with the highest fidelity possible with current technology (see Figure 13). The key to Jules is two innovations. The first is 'frubber' (short for 'flesh rubber'), a patented spongy elastic polymer that provides a convincing emulation of the way human flesh, and the muscles underneath, moves. The second innovation is the design and linkage, via wires that push and pull the light frubber skin from behind, of thirty-two servomotors.

When actuated in the correct combination and with the correct timing, these motors provide realistic emulation of about fifty muscle groups in the human face, thus allowing Jules a wide range of facial expressions, from frowning to smiling, puzzlement to dismay. Jules is intended to provide a face-to-face conversational capability between robot and human, and Jules's lips and mouth are designed to be actuated in synchrony with computer-synthesized speech, and do, indeed, provide a reasonable emulation of the way human lips move during speech.

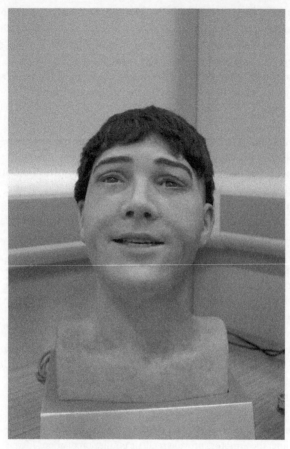

13. Jules, an expressive high-fidelity android robot head

Jules also has motors for gross head movements to nod or tilt
the head, or simply provide small random movements (a robot
head that is completely still would look unnatural). Of even
greater importance are Jules's eyes, which not only look
human but are actuated to saccade in a lifelike way; furthermore,

the eyes contain cameras, so that with appropriate face recognition and tracking software, linked to the eye motors, Jules can look at and track a human face. To complete the illusion, the robot's eyelids are actuated to provide Jules with a periodic blink reflex.

An interesting and perhaps unique aspect of Hanson's android robot heads is that the robot's facemask can be modelled on a real person, and indeed one famous example is the Hanson robot model of science fiction author Philip K. Dick. With customized conversational software based on chatbot technology (which I discuss in Chapter 6), one could (somewhat weirdly) have a conversation with a robot that not only looks like a real person but appears to have the memories or personality traits of that person. Jules, however, is not modelled on a real person but was designed with statistically determined androgynous features.

There are clearly ethical questions raised by the fashioning of robots that behave *as if* they have feelings, since—whether the robot is a cartoon humanoid like Kismet or a high-fidelity and roid like Jules—the emotional states are entirely in the eye of the human beholder. None of these robots, nor any robot yet built, can truly be said to have feelings. They are at best robot actors, designed to deceive—and perhaps to shock if they descend into 'the Uncanny Valley' (see Box 2).

What if the robots are designed not for teenagers or adults fully cognizant of the artificiality of the robot and its emotions, but for children, naive, or vulnerable users who would have difficulty understanding that it's a robot pretending to be happy, or sad, or pleased to see them? I believe this to be a genuine concern and something that requires ethical guidelines, as I discuss later in this chapter.

Box 2 The Uncanny Valley

Builders of humanoid robots face a unique and fascinating problem, known as the Uncanny Valley. Proposed by roboticist Masahiro Mori in 1970, the Uncanny Valley imagines how humans might react as robots become more human-like. His graph, shown here in Figure 14, has a horizontal axis that runs from 0 per cent human-like on the left to 100 per cent human-like on the right.

Mori argued that our reaction to robots will become more positive the more human-like the robot is, until the humanoid robot is very close in appearance and behaviour to real humans; but then, dramatically, we will have a strong adverse reaction to near-human-like robots. It is this reaction that Mori termed the Uncanny Valley. Mori also suggested that this adverse reaction will be more extreme when the robots are moving than when they are static. It is an idea that makes reference to our extraordinary sensitivity to appearance or behaviour in each other that differs—even in very subtle ways—from the norm.

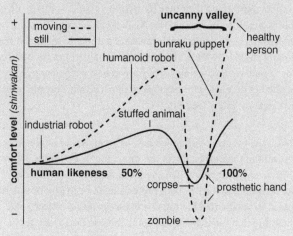

14. The Uncanny Valley

Is the Uncanny Valley real and if so does it pose a serious problem in the development and use of humanoid robots? Certainly my own experience of the reaction of visitors to the lab to one of David Hanson's android robot heads—particularly fright in young children—does seem to bear out the Uncanny Valley hypothesis (although the fact that the bodiless robot heads are just sitting on a table top may have something to do with the children's shock). Roboticist Hiroshi Ishiguro has carried out studies in which human subjects were asked for their reaction to computer-generated images of heads that 'morph' from humanoid to human—and these studies do appear to confirm Mori's hypothesis.

What are the consequences of the Uncanny Valley? On the face of it, it would suggest that there's no point building realistic android robots until some future time that we are able to make them so realistic in both behaviour and appearance that they climb out of the Uncanny Valley. But, for some applications, robots that remain in the Uncanny Valley may have a role: for instance, humanoid robots required to behave only as animated mannequins—such as automated receptionists in high-tech companies—or in roles that would benefit from the uncanny such as the arts or theatre. Does the Uncanny Valley rule out android robots from roles in which they are required to interact with children, the elderly, or vulnerable people? Perhaps. It may be that these groups would be better served with either cartoon robots or humanoid but not android robots. But it's hard to speculate. We don't yet know enough about the psychology of human–robot interaction.

Robot pets: zoomorphic robots

Robots whose shape and structure are modelled on animals are referred to as zoomorphic, and many of the reasons for building or using humanoid robots apply equally to zoomorphic robots. Robot pets, for example, are designed as toys or companions for people.

Perhaps the most famous example is the Sony Aibo robot dog, while a rather less sophisticated—but huggable—robot pet is the Paro robot seal-pup from another Japanese company, Intelligent Systems Co. (see Figure 15).

Paro is an advanced zoomorphic robot modelled on the baby harp seal. Designed specifically to provide comfort and as a therapeutic aid in hospitals or elder care homes, the robot has five types of sensor, for touch, light, sound, temperature, and orientation. The robot is actuated for a range of head movements, blinking eyelids, front leg movements, and tail movements. The robot's AI is programmed so that Paro will respond to touch or strokes by both moving its body and making baby seal 'cries', and a degree of randomness in the robot's responses provides a convincing and comforting animation.

Notably, Paro is able to learn: if it is stroked, then it is more likely to repeat its behaviour prior to being stroked. If the robot is

15. Paro, a robot baby seal

instead hit, then it is less likely to repeat the previous action, thus providing the robot with simple Skinnerian conditioning. The robot can also learn its name, when spoken, and is programmed to turn to the direction of a voice, and has a range of artificial emotional responses, such as surprise, happiness, and anger. Paro has been the subject of a number of documented trials and does indeed appear to be effective as a surrogate pet. In my view Paro is an important development in the early evolution of robot companions.

Sex robots

An obvious market for humanoid robots that we're likely to see in the near future is as sex toys. The sex industry is well known as an early adopter of new technology; one only has to think of the videocassette and the Internet. It comes as no surprise, therefore, to learn that sex entrepreneurs are already fitting sensors, actuators, and simple artificial intelligence to sex dolls. The robot Gigolo Joe, played by Jude Law in the movie *A.I. Artificial Intelligence*, remains a distant fantasy, but it seems depressingly possible that the first mass market for humanoid robots might be as sex toys or sex companions.

While this may appear harmless for adults, there are, I think, two dangers, one short term and one longer term. The near-term danger is safety; to put it bluntly, sex with a robot might not be safe. As soon as a robot has motors and moving parts, then assuring the safety of human–robot interaction becomes a difficult problem and if that interaction is intimate, the consequences of a mechanical or control systems failure could be serious.

In the longer term the dangers are of a different kind. If a sex robot is not much more intelligent than the washing machine, then it poses no real moral hazard, but imagine a much higher level of machine intelligence in a sex or companion robot. The possibility of intimate human–robot relationships raises a wide range of difficult societal, psychological, and ethical issues. Some have written of the possibility of humans falling in love with humanoid robots and sustaining long-term, even marital, relationships.

While the subject of love and marriage with robots is not something that I consider here, it does open up the more interesting question of the authenticity of a robot's emotional responses. Many would argue that a robot that only ever behaves *as if* it has feelings could never be a genuine companion at an emotional level. I disagree—I believe that if you can trust a robot, then, for a sufficiently advanced robot, the 'as if' argument disappears.

Cronos—a case study for future humanoid robots

Perhaps one of the strangest humanoid robots built to date is Cronos (see Figure 16). With its skeletal appearance and single eye, Cronos might have been designed for a horror movie, but the robot's appearance is simply a consequence of the radical ideas behind it—ideas that could have far-reaching consequences for the way we build humanoid robots, and for their AI.

Cronos's body is formed from a set of hand-made structural elements literally hand-sculpted from a low-temperature thermo-softening plastic. While the structural elements of the 'skeleton' are not all intended to be anatomically accurate, many were copied from *Gray's Anatomy* and look uncannily like real bones—this is not too surprising when the overall objective is a truly humanoid robot. Where Cronos departs from conventional robot design is that the resting position of the joints between individual 'bones' is not rigidly set by the position of an electric motor but by a series of carefully counterbalanced springs and tendons, many of which are made from elastic (bungee) cords. This means that Cronos's body is highly compliant and elastic: push against one of its arms and not only will the arm yield to your push, but the whole body will flex in response.

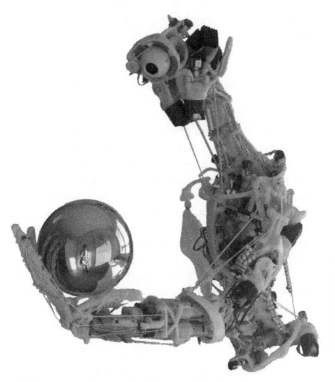

16. Cronos, an anthropomimetic humanoid robot

Cronos does have motors, of course, but these, positioned to correspond to human muscles, also act by pulling on elastic cords connected to the skeleton so that a single motor movement causes the whole robot to move in a complex way. At the end of the movement, the robot will typically 'bounce' before settling into its new position. This is in marked contrast with traditional rigid robot design approaches (which industrial robot arms and humanoid robots like BERT have in common), where great efforts are made to ensure that each single joint motor movement affects only the position of that one particular joint; the advantage of the traditional 'stiff' approach is that robots are (relatively) easy to control. Because of its compliance, Cronos is very difficult to control—but in fact this is the whole idea.

Cronos was conceived within a project to explore the (hard) question of machine awareness. Roboticist Owen Holland and his colleagues reasoned that a complex and hard-to-control body requires a new approach to how a robot controls its own body. Traditional stiff robots like BERT can have body movements specified more or less exactly by sets of equations and then programmed as control algorithms. This approach is impossible with a robot like Cronos, and instead the robot has, within itself, a computer simulation of itself. Cronos then uses its internal model in the following way: when faced with a task Cronos first sets up its virtual model of itself and its environment so that it matches the physical situation at the start of the task (e.g. an object on a table that must be knocked over), then tries out different motor programs on its virtual self until it finds one that achieves the task (i.e. the object has been successfully knocked over). It's as if the robot has an imagination and is able to rehearse its movements and interactions before trying them out for real.

Ideally, Cronos should be able to learn its own body plan and dynamics—in effect, to build its own internal model of itself—then continuously adapt and update that internal model as the robot operates and as its joints and tendons wear, for example. To achieve this, Cronos would need to act as follows: during a developmental phase the robot would try out different moves, in effect exercising its own muscles and joints. The actual effects of those movements on Cronos's body would be measured and fed back into the robot simulator—its internal model of itself—so that over time the model improved. Thus, in a process that might be likened to body babbling in infants, Cronos would learn how to control its own body. This idea is now being studied in ECCERO-BOT, Cronos's even more complex successor.

In Chapter 2, I introduced the idea of biomimetic robots, and all humanoid robots are, of course, biomimetic in the sense that they emulate the general human shape and form. Cronos goes much further in that it is anthropomimetic, i.e. it mimics the human form on the inside as well as the outside. This means, argue Holland and his colleagues, that the way Cronos is controlled must be a better emulation of the way we humans control ourselves and our physical interactions with the world. They hope that because of its design, Cronos and its successors will demonstrate some human-like cognitive features.

The particular significance of Cronos is that it points the way to a future generation of humanoid robots that are light, elastic, and compliant. These qualities will be essential for robots that closely interact with humans, because they minimize the likelihood that the robot can cause physical harm to a human. If a compliant robot's arm should accidentally impact part of your body, then its natural compliance will absorb most of the energy of impact—the robot's arm and body will 'give'—and you will not be hurt.

Of course, the robot's sensors and AI should also be designed to minimize the possibility of an accidental impact, but guaranteeing the safe operation of intelligent control in unpredictable environments is more or less impossible; thus the robot's natural compliance is the last resort in case its other safety features fail. Future humanoid robots will need to be light and compliant because only that way can they be safe.

The safety and trustworthiness of human-robot interaction

All of the potential applications of humanoid robots, which I have broadly divided into the robot *workplace assistant* or the robot *companion* (including conversational, therapeutic, and zoomorphic robots), have one thing in common: close interaction between human and robot. The nature of that interaction will be characterized by close proximity and communication via natural human interfaces—speech, gesture, and body language.

Human and robot may or may not need to come into physical contact, but even when direct contact is not required they will still need to be within each other's body space. It follows that robot safety, dependability, and trustworthiness are major issues for the robot designer. But, given that we humans are unpredictable, then how can we design and build human robots to be safe in all circumstances?

Robots, like any machine that is tasked or entrusted with a particular job, need to be designed to be safe and reliable. This is the same level of dependability we would expect from our car or washing machine, i.e. that it's been well designed, and built to meet or exceed standards of manufacture and product safety.

But making a robot safe isn't the same as making it trustworthy. One person trusts another if, generally speaking, that person is reliable and does what they say they will. So if I were to provide a robot that helps to look after your grandmother and I claim that it is perfectly safe—that it's been designed to cover every risk or hazard—would you trust it? The answer is probably not.

Trust in robots, just as in humans, has to be earned. First, you would like to see the robot in action (preferably not with your grandmother). Perhaps you would like to interact with it yourself; in so doing you build a mental model of how the robot behaves and reacts and, over time, if those actions and reactions are consistent and predictable for the circumstances, then you will build a level of trust for the robot. The important thing here is that trustworthiness cannot just be designed into the robot—it has to be earned by use and by experience.

Consider a robot intended to fetch drinks for an elderly person. Imagine that the person calls for a glass of water. The robot then needs to fetch the drink, which may well require the robot to find a glass and fill it with water. Those tasks require sensing, dexterity, and physical manipulation, but they are problems that can be solved with current technology.

The problem of trust arises when the robot brings the glass of water to the human. How does the robot give the glass to the human? If the robot has an arm so that it can hold out the glass in the same way a human would, how would the robot know when to let go? The robot clearly needs sensors in order to see and feel when the human has taken hold of the glass.

The physical process of a robot handing something to a person is fraught with difficulty. Imagine, for instance, that the robot holds out its arm with the glass but the human can't reach the glass. How does the robot *decide* where and how far it would be safe to bring its arm toward the person? What if the human takes hold of the glass but then the glass slips; does the robot let it fall or should it—as a human would—renew its grip on the glass?

At what point would the robot decide the transaction has failed: it can't give the glass of water to the person, or they won't take it; perhaps they are asleep, or simply forgotten they wanted a glass of water, or confused. How does the robot sense that it should give up and perhaps call for assistance? These are difficult problems in robot cognition. Until they are solved, it's doubtful we could trust a robot sufficiently well to do even a seemingly simple thing like handing over a glass of water. So how might we begin to consider designing a robot that would be trusted with this kind of task?

From a technical point of view, the robot needs two control systems: one is the cognitive system that actually carries out the task. Another, parallel, *safety system* is one that would constantly check for unexpected faults or hazards. The primary job of the safety protection system is to stop the robot, but in a safe fashion (noting that there are some situations where freezing the robot would itself be an unsafe thing to do). First we must solve the problems of cognition and safety. Next a robot must prove itself dependable in use. Only then is it likely to earn our trust.

Robot ethics

In his 1942 short story, 'Runaround', Isaac Asimov famously put forward his three laws of robotics and in so doing introduced the idea that robots could or should behave ethically. (As an aside, Asimov was also the first to coin the term 'robotics'.) Asimov's laws of robotics were, of course, a fictional device. Asimov himself never seriously expected that future roboticists would contemplate building them into real robots—he was well aware of how difficult this would be.

But his idea that robots should be 'three laws safe' has become part of the robotics discourse. It is hard to debate robot ethics without acknowledging Asimov's contribution, and rightly so. Let us remind ourselves of his three laws of robotics: first, a robot may not injure a human being or, through inaction, allow a human being to come to harm; second, a robot must obey any orders given to it by human beings, except where such orders would conflict with the first law; and third, a robot must protect its own existence as long as such protection does not conflict with the first or second law. Asimov later added a fourth law: a robot may not harm humanity, or, by inaction, allow humanity to come to harm—which came to be known as the zeroth law, since logically, it precedes the first three.

The fundamental problem with Asimov's laws of robotics, or any similar construction, is that they require the robot to make judgements. In other words, they assume that the robot is capable of some level of moral agency. To see why this is so, consider the first law: may not injure… or, *through inaction*, allow a human being to come to harm. 'Through inaction' implies that a robot is capable of determining that a human is at risk, and able to decide if and what action is needed to prevent the possible harm. The second and third laws compound the problem by requiring a robot to make a judgement about whether obeying a human, or protecting itself, conflicts with the first (and second) law. No robot that we can currently build, or will build in the foreseeable future, is 'intelligent' enough to be able to even recognize, let alone make, these kinds of choices.

Of course, even if a far-future robot were intelligent enough to make moral judgements, for it to be allowed to do so, society would have to grant it the right to be regarded as a moral agent—in other words, grant the robot personhood (or something very much like it). And with rights come responsibilities, which would again pose a difficult problem to society: what sanctions, for instance, would be appropriate for a robot that broke the robot laws? For robots to be 'three laws safe' would require not only very significant advances in robot AI, but also a huge change in robots' legal status, from products to moral agents with rights and responsibilities.

Most roboticists agree that for the foreseeable future robots cannot be ethical, moral agents. (Although some, controversially, argue that near-future robots that have an artificial conscience—so that the robot's behaviours encode the military rules of engagement—could be built.) However, roboticists are also agreed that robot ethics is an important issue for the community, and several serious efforts have been made in this direction, including a European *Roboethics Roadmap*, a South Korean *Robot Ethics Charter*, and in Japan draft *Guidelines to Secure the Safe Performance of Next Generation Robots*.

So why, if robots cannot be ethical, is robot ethics an issue? There are two reasons. The first is that robots are beginning to find application in human living and working environments, and — as I've described in this chapter — those robots will be required to interact closely with ordinary people, including children, vulnerable, or elderly people: applications that require strong safeguards regarding design, operation, and privacy.

The second is that precisely because, as we have seen, present-day 'intelligent' robots are not very intelligent, there is a danger of a gap between what robot users believe those robots to be capable of and what they are actually capable of. Given humans' propensity to anthropomorphize and form emotional attachments to machines, there is clearly a danger that such vulnerabilities could be either unwittingly or deliberately exploited.

Although robots cannot be ethical, roboticists should be. Robotics researchers, designers, manufacturers, suppliers, and maintainers should, I advocate, be subject to a code of practice, with the founding principle that robots have the potential for very great benefit to society. What might that code of practice look like?

One set of draft ethical principles for robotics, drafted by a UK working party, proposes:

1. Robots are multi-use tools. Robots should not be designed solely or primarily to kill or harm humans, except in the interests of national security.
2. Humans, not robots, are responsible agents. Robots should be designed and operated as far as is practicable to comply with existing laws and fundamental rights and freedoms, including privacy.
3. Robots are products. They should be designed using processes which assure their safety and security.
4. Robots are manufactured artefacts. They should not be designed in a deceptive way to exploit vulnerable users; instead their machine nature should be transparent.
5. The person with legal responsibility for a robot should be attributed.

Importantly, these five draft principles downplay the specialness of robots, stressing instead that robots are tools, products, and artefacts that have to be designed and operated within legal and standards' frameworks. Only the fourth explicitly addresses a quality unique to (some) robots, that of creating emotional bonds or dependencies of human on robot, by proposing that designers or manufacturers should not exploit such dependencies and that it should always be possible, like Toto in *The Wizard of Oz*, to pull aside the curtain and expose the robot underneath.

Chapter 5
Robot swarms, evolution, and symbiosis

Swarm robotics

When instead of a single robot we use a system of multiple robots working together to achieve a task, we are in the domain of multi-robot systems. Although the idea of multi-robot systems is not new, it is only in recent years that commercial systems have become available. A very good example is the Kiva warehouse robot system, described by the company as a 'goods-to-man order picking and fulfilment system'.

Kiva robots are strong battery-powered wheeled mobile robots about 60 cm wide, 76 cm long, and 30 cm high. A distinctive orange colour, each Kiva robot has, on its top surface, a metal lifting plate; the robot positions itself underneath individual person-height storage shelving units, then physically lifts the storage unit a few centimetres off the floor. The lifting action is interesting. The lifting mechanism uses a screw drive, which rotates the lifting plate (imagine unscrewing a screw), but while this is happening the robot pirouettes in the opposite direction at the same speed. This has the effect of keeping the storage unit straight and stable while it's being lifted.

Individual Kiva robots navigate around the warehouse using a grid system of 2D barcodes on the floor, and the whole multi-robot fleet is choreographed by a central computer system that keeps track of where the robots are, where the inventory is, and where it needs to go. The robots, collectively, fetch the different items from the warehouse to a human, who takes them off the shelving units and makes up the complete order. Shelving units are then returned to the warehouse. And interestingly, over time, the system optimizes the position of the shelving units in the warehouse, so that more popular items end up close to the packing station—minimizing time and energy costs. It's an ingenious system that represents the state of the art in real-world multi-robot systems.

The Kiva robot system is an example of a multi-robot system with centralized control and coordination. Swarm robotics offers an alternative approach in which there is no centralized control. Instead control is decentralized and distributed between the robots themselves. It is an approach that embodies the principles of swarm intelligence.

Swarm intelligence

Think of a termite mound. It is an astonishing construction, not only for its size relative to its tiny blind constructors, but also for the sophistication of its architecture, with temperature and humidity control, fungus gardens, and nursery chambers. The puzzle is how animals that appear individually to be so incapable, and (relatively) unintelligent, can organize and coordinate themselves to build, maintain, and repair this remarkable structure. We humans would require an architect to design the structure, then a hierarchy of planners, civil engineers, foremen, and labourers, each instructed to carry out a carefully coordinated task, in the right place and at the right time, in the overall construction.

Yet we know that the termites have no such hierarchy. There is no 'brain' termite that instructs each worker on the task it must carry out. Instead each individual termite has a repertoire of simple instinctive behaviours, each one of which is triggered by some condition; this may be a chemical (pheromone), or interaction with another termite, or physical contact with some part of the nest. No individual has a plan of how to construct the nest, so no matter how industrious, it could not build one on its own. But collectively, thousands of termites do achieve this remarkable feat of civil engineering.

The termite mound, its construction, and maintenance are said to be an *emergent* property of the very large number of micro-interactions of termites with each other and with their physical environment. The kind of self-organized collective intelligence demonstrated by termites and social insects generally is called *swarm intelligence*, and multi-robot systems based on the same principles are called swarm robotic systems.

The defining characteristics of swarm robotic systems are: first, local sensing—each robot senses only its local environment; second, local communications—each robot communicates only locally with its nearest neighbours; and third, autonomy—based only on local sensing and communications, each robot decides for itself which action to take.

Often the robots in a swarm are all the same and—importantly— run the same control program. In other words, they have an identical set of built-in behaviours. Thus, in a swarm robot system there is no hierarchy, no predetermined leader (although one might emerge, perhaps temporarily), and control is completely distributed between the robots of the swarm.

In robotics, swarm systems hold a particular fascination. From a scientific point of view they are interesting as working models for swarm intelligence, and it's not surprising that some biologists work with roboticists to try and discover how insect 'societies' work: a good example of a problem studied with robots is adaptive division of labour in ant colony foraging. For a given amount of 'food' in the environment, how does the swarm automatically balance the number of robots out collecting food against the number of robots that stay in the nest conserving energy, in order to optimize the overall swarm energy? (If food is scarce and all robots are out foraging—an activity that consumes a lot of energy—the swarm would run out of energy; if, on the other hand, all robots stay in the nest to save energy, the colony will eventually starve.) Not an easy problem to solve when individual robots have no global knowledge of how much food is available to collect, or the overall energy level of the swarm.

In swarm intelligence, not for the first time in robotics, engineering and biology strongly intersect. But from an engineering point of view a robot swarm is also extremely attractive. One reason is robustness. A swarm of robots has a very high tolerance to failure. If some of the robots in the swarm fail, the remaining robots will, in principle, complete the task. It may take longer, but it will still get done. Another reason for a swarm's robustness is that there is no central control computer, or its communications network with the robots, to go wrong. The swarm approach is attractive also because of its scalability. Imagine a swarm of robots performing a clean-up operation. In principle, if you want the job done in less time, simply add more robots. (This will only work if there's enough space for the robots to work in; if they start to get in each other's way, the advantage of adding extra robots will be lost.)

Potential real-world applications

Many real-world applications of robotics are for tasks that are spatially distributed. Think of robotic exploration, survey, environmental monitoring, or search and rescue; all would require robots to cover a large physical area and, clearly, a number of robots that are distributed across that area could between them cover the territory in much less time than one. The same is true for agricultural or horticultural robots, and for industrial cleaning or waste collection robots. Consider also future medical micro-robots that might be injected into the bloodstream; almost certainly the task would require many, perhaps thousands, of microscopic robots. All of these are potential applications for swarm robotic systems.

In fact, there is a very useful metaphor for all of these potential applications of collective robotics, and that is *foraging*. We think of foraging as the collection of food, but the idea can be generalized to searching for, collecting, and retrieving anything, including information. Foraging has therefore become a benchmark problem in swarm robotics, and foraging algorithms could be applied, with appropriate modification to fit the actual task, to any of the real-world applications listed above.

Case study: Swarm-bot

The Swarm-bot project, led by roboticist Marco Dorigo, provided a fascinating laboratory demonstration of swarm robotics in a search-and-rescue scenario. The individual robots, called s-bots, are searching for survivors. One or two s-bots find a child. The s-bots are equipped with a gripper to grab the child's clothing, but on their own the robots are too small and light—about 12 cm in diameter and 660g in weight—to be able to drag her to safety. So the s-bots that found the child and grabbed her clothing signal for others to join them. Soon other s-bots from the swarm reach the child, but then there's another problem. Given the relative sizes of the child and the s-bots, there's room for only four robots to be able to grab the child's clothing with their grippers. Even four s-bots are not strong enough to be able to pull the child, so the remaining free robots use their grippers to grab the robots that are already holding her, and soon they self-organize into four robot chains. After a while each robot chain is long enough to have enough traction and the child is physically dragged to safety. The swarm of s-bots is collectively known as Swarm-bot.

The child-pulling Swarm-bot illustrates a number of important characteristics of swarm robot systems. First, they typically need more than one individual robot to complete the task. In the demonstration I have outlined here, eighteen s-bots are needed. If eighteen is the minimum needed, then the failure of even one s-bot would compromise the task. Swarm robot systems normally employ more than the minimum number of robots to provide a margin for some robots failing. But this redundancy is not a problem. In fact, it's an advantage. If, for example, the child-pulling example had twenty-four instead of eighteen robots, collectively it would be able to search faster, and rescue her in less time, yet still be able to tolerate the failure of up to six robots. It would be extremely difficult to design a single robot to be able to tolerate an equivalent 25 per cent failure of its subsystems.

Second, we see that the robots in the swarm need to be able to cooperate and coordinate their actions and work as a team. In this example, the robots must communicate with each other so that they can agree when there are enough to be able to pull the child, then synchronize so they all start pulling at the same time and in the same direction.

This example illustrates another advantage of swarm robot systems, which is that the individual robots in the system are typically smaller (and perhaps simpler) than a single robot would have to be to accomplish the same task. Imagine how large a single robot would have to be to rescue a child. Smaller, simpler robots are typically each more reliable than the equivalent same-task-achieving single robot.

Case study: Swarmanoid

The Swarm-bot case study above is an example of a homogeneous system, because each of the individual robots (s-bots) is identical. But the robots in a swarm don't have to be identical. In a heterogeneous system we would find several different types of robot, each with a different functional specialism. This gives the overall system much more flexibility to undertake a wider range of tasks, while retaining the robustness of the swarm combined with the relative simplicity—and hence reliability—of the individual robots.

Figure 17 shows the three different types of robots that comprise Swarmanoid, called the Foot-bot, the Hand-bot, and the Eye-bot. Also led by Marco Dorigo, the Swarmanoid project is a successor to Swarm-bot, and not surprisingly the Foot-bot is a development of the s-bots I described earlier. The Foot-bots are ground based and able to move both individually and, when linked together with their grippers, as a team. But of course they are limited to operating in 2D, rather close to the ground.

The Hand-bot, however, is specialized for gripping objects. It has no wheels of its own and must be carried by Foot-bots, as shown in Figure 17. However, the Hand-bot is not limited to operating close to the ground. It has a device — called the ceiling attachment system — which enables it to project a cord to the ceiling, then climb up the cord to grab objects between the floor and the ceiling. While suspended from its cord, the Hand-bot can control its orientation with two fans.

17. The Swarmanoid robots. Left: three Foot-bots and a Hand-bot. Right: the Eye-bot

The third robot, the Eye-bot, is a quadrotor flying robot. It can therefore move in 3D space and its function is to provide an 'eye in the sky' for the overall system, thus allowing the Foot- and Hand-bots to coordinate their actions and work as a team. Like the Hand-bot, the Eye-bot also has a ceiling attachment device which means it can 'park' on the ceiling to provide a fixed stable position for its camera, while conserving battery power. In what was almost certainly the world's first demonstration of a heterogeneous swarm system collaborating to achieve a complex 3D task, the Eye-bot coordinated a group of Foot-bots and a Hand-bot to locate and fetch a book from a high bookshelf.

So far I've highlighted the advantages of swarm robot systems, but what are the downsides? The first, not so much a downside but a simple consequence of the obvious complexity of collective robot systems, is that they can be very challenging to design. Even if a collective robot system has only one kind of robot, as in the child-pulling Swarm-bot example, the design of the systems that coordinate the robots so that, collectively, they will carry out the required tasks presents particular difficulty (and therefore interest).

Another downside is the problem of how humans command and monitor the swarm. Swarm robotic systems are normally designed with a high level of autonomy, but even highly autonomous robot systems need a human interface. At the very least this will allow human operators to both command the swarm to achieve some high-level objectives and monitor its progress—with the option of either changing the mission objectives or aborting the mission altogether if something goes wrong.

Evolutionary robotics

One of the most fascinating developments in robotics research in the last twenty years is evolutionary robotics. Evolutionary robotics represents nothing less than a new way of designing robots. It uses an automated process based on Darwinian artificial

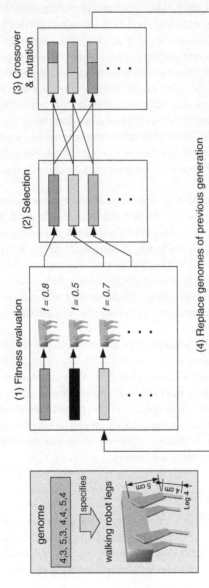

18. The four-stage process of evolutionary robotics

selection to create new robot designs. Of course, selective breeding, as practised in human agriculture to create new improved varieties of crops, or farm animals, is (at least for now) impossible for real robots. Instead, evolutionary robotics makes use of an abstract version of artificial selection in which most of the process occurs within a computer. This abstract process is called a genetic algorithm.

The process of evolutionary robotics works as follows. We first represent the robot that we want to evolve with an artificial genome. Rather like DNA, our artificial genome contains a sequence of symbols but, unlike DNA, each symbol directly represents (or 'codes for') some part of the robot. In evolutionary robotics we rarely evolve every single part of the robot.

Imagine how we might construct a genome for a four-legged walking robot. Each leg has two segments and the genome specifies the length of each segment. Since there are four legs, the genome contains eight values in total. The box at the left of Figure 18 shows the genome specifying a walking robot's legs.

Of course, there are many other dimensions of the robot that we could also code for in the artificial genome. We could, for instance, code the size of the robot's torso in the genome, or the positions of the joints between the torso and the legs, or the number of segments in each leg, and so on. In theory we could code for every single aspect of the robot in the genome but that would not only slow down the artificial evolutionary process, but also reduce the likelihood of us evolving a 'good' robot. In practice therefore, we hand-design and fix some aspects of the robot, decide which parts of it we want to evolve, and code only those parts in the artificial genome.

Having determined the structure of the artificial genome, we now create a population of genomes in a computer program. It doesn't need to be a large population; between ten and a hundred individuals is sufficient to maintain some genetic diversity. The individuals in the initial population are, as a rule, randomly initialized. For our imaginary four-legged walking robot this would mean that the initial population would consist of robots with mismatching leg sizes, many of which would not be successful walkers.

Figure 18 illustrates the overall four-stage process of artificial evolution. For each population, each individual in the population—represented by its genome—is subjected to a separate fitness evaluation. This is labelled as (1) 'Fitness evaluation' in Figure 18. What this means is that each genome is used to create a robot. It would be hugely impractical to create a real physical robot for each genome, so in practice we create a simulated robot inside a computer.

That simulated robot is then tested. In other words, its performance is measured against some fitness criteria. For our four-legged walker, a successful robot would be able to walk in a straight line—the *straighter* the line and *faster* the walking speed, the more successful (i.e. 'fitter') the robot. After the fitness trials, each individual in the population has a fitness value. In Figure 18 these are shown as '$f = 0.8$', etc. A higher value means a higher fitness.

Those fitness values are then used to select which individuals are to reproduce, as illustrated by stage (2) 'Selection' in Figure 18. Here the two individuals with $f = 0.8$ and $f = 0.7$ are selected. The individual with a low $f = 0.5$ is not selected. There are many selection strategies that we could use here. We might, for instance, simply choose the fittest 50 per cent of the population. This half then generates the next population.

The next generation is then created, as illustrated in stage (3) 'Crossover and mutation' in Figure 18. Again there are many ways in which we can do this (remember, these are just numbers in a computer); one would be to reproduce asexually, thus for every successful individual we could generate two copies, each one slightly mutated. Here mutation simply means that each (or some) of the values in the artificial genome is altered by some small, randomly chosen amount.

Another approach is to take pairs of fit individuals from stage (2) and swap parts of their genomes to generate offspring, as shown in Figure 18. Each new genome contains two parts, one part from each 'parent' genome. The swapping process is called crossover and in some very abstract sense this is analogous to sexual reproduction. Whatever strategy is chosen for selection, reproduction, and mutation, the next generation (which maintains the same population size) is created and this new set of genomes replaces the genomes of the previous generation, in stage (4) of Figure 18.

The individuals ('children') in this new population are subjected, in exactly the same way and using the same fitness criteria, to fitness trials. This whole process is then repeated, quite possibly over hundreds of generations, until we have evolved a robot that successfully achieves the desired performance. In other words, we stop after stage (1) when one individual genome generates a robot with a fitness that exceeds some chosen value of f. Then we construct a real-world version of that winning robot design and the process is complete.

Evolving robot bodies

In the landmark Golem project, Hod Lipson and Jordan Pollack evolved strange new robots, capable of moving themselves across a flat surface. Figure 19 shows one of the evolved robots, on the top the most successful virtual robot in its virtual world, and on the bottom the real robot fabricated with a 3D printer to the same specification, with motors added.

Unlike with our imaginary four-legged walking robot, Lipson and Pollack allowed the evolutionary process much more freedom to evolve the robot's body shape. They determined only

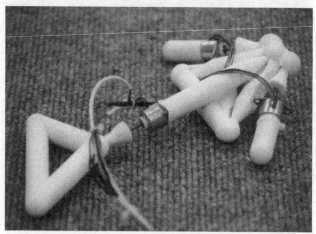

19. In the Golem project, a robot is evolved in simulation (top), then the final best individual is fabricated (bottom)

that robots would be made from a number of segments, some of which have a fixed length, and others that are able to telescopically shrink or expand along their length. These moving segments shrink and expand at a fixed rate and act as a kind of continuously flexing-unflexing muscle for the robot. The evolutionary process was free to vary the number of segments, the length of the segments, and where—and at what angle—the segments are joined. The fitness evaluation was based simply on how far the robots could move (if indeed they moved at all) from a starting position in a fixed time. Those that moved the farthest were selected and became the parents for the next generation.

The Golem project successfully evolved a number of robots with the ability to physically move themselves across the surface. All had strange and sometimes surprising body shapes: almost certainly not designs that a human engineer would come up with. The genetic algorithm discovered equally unexpected ways of achieving locomotion, exploiting both the movement of the 'muscle' segments and the friction between the body segments in contact with the floor, and the floor itself. Locomotion emerges from the rather complex interaction between the robot's body parts and the floor surface, subject to the physics of mass and friction—so it would be very difficult to predict how these robots move without trying them out in the simulated world.

Also surprising was that some of the evolved robot bodies, including the one in Figure 19, are — like most animals — symmetrical. Symmetry was not rewarded by the fitness evaluation, so the emergence of symmetry probably reflects the fact that symmetrical robots are more likely to move in a straight line and therefore travel farther from the start position than asymmetrical ones. So the bilateral symmetry we see here — having a left-hand side that mirrors the right — was just a side effect of evolution. It's easy to imagine creatures a bit like this crawling across the Cambrian sea floor.

Evolving robot brains

In the Golem project, the whole evolutionary process was carried out in a computer, including fitness evaluation of simulated robots in a virtual world. Although successful in that project, the simulation-based approach to evolving robots is challenging. The biggest problem is that the simulator used to carry out the fitness trials needs to be able to faithfully simulate the characteristics of each of the parts of the robot and the experimental world. As I've stressed throughout this book, given that the behaviour of a robot is the result of the interactions between the robot and its world, then those interactions must be realistically simulated (see Box 3).

If there is a mismatch between any aspect of the simulated robot and its test arena, and the real robot and real test arena, then the evolved robot is unlikely to work as well as it appeared to work in simulation when it is created and tested for real. This problem, which is known in evolutionary robotics as 'the reality gap', is difficult but not insoluble. The advantages of the simulation-based approach are sufficiently great that a good deal of effort and ingenuity has gone into ways of working around the reality gap.

The Golem project evolved robot bodies, but not robot 'brains', in the sense that the robot's control systems were fixed—a continuous cyclic flexing of the robot's 'muscle' segments. If we want to evolve the robot's control system, the problem of the reality gap becomes more acute, and evolutionary roboticists have adopted a different approach.

This approach is to hand-design and therefore fix the design of the robot's physical body altogether and evolve only the robot's control system. Here the genome codes only for the robot's control system, which means that all fitness trials can be conducted using the same physical robot; in fact, just one physical robot. The process remains essentially the same as shown in Figure 18. The only difference is that in stage (1) 'Fitness evaluation' each genome is used to create a new control system, which is then downloaded into a real robot in a real test arena. This has to be done for each genome in each generation, so it's a time-consuming process. A population size of twenty robots, evolved for say one hundred generations, will require 2,000 separate fitness trials. But the advantage is that the fitness values f will be based on the real robot in the real world, so completely avoiding the problem of the reality gap.

In a very elegant example of this approach, Markus Weibel, Dario Floreano, and Laurent Keller used the miniature sugar-cube-sized robot ALICE to evolve robots that demonstrate cooperative and altruistic behaviours. Figure 20 shows both the real robot and its control system.

In this experiment the robot controller is an artificial neural network (ANN), shown in Figure 20 (B). An artificial neural network is a highly simplified model of biological neurons. Each artificial neuron has multiple inputs and a single output; the output of the artificial neuron is a weighted sum of the values of the inputs, and the weights represent the 'strengths' of the inputs to a neuron. (The 'weight' is a multiplier value. Thus if the input value is 2, and the weight is 0.5, the weighted value is $2 \times 0.5 = 1$.) ANNs are important in robotics and of special value to evolutionary robotics, because we can code the weights of each input to each neuron in the artificial genome, then evolve, in the way I have described, the robot's ANN.

Figure 20 shows how the ANN is wired to the input sensors and output motors of the ALICE robot. In this example the ANN has a total of six neurons, shown as light grey circles. The inputs on the left-hand side—the seven small dark circles—are connected directly to the robot's sensors, in this case four infrared sensors and two cameras. Two of the output neurons, on the right-hand side, are connected directly to the left- and right-hand motors of the robot. Each of the black connecting lines is a weighted input to a neuron, and it is these weights that are evolved.

Compare Figure 20 with Figure 2 in Chapter 1 and we see that the artificial neural network has completely replaced the

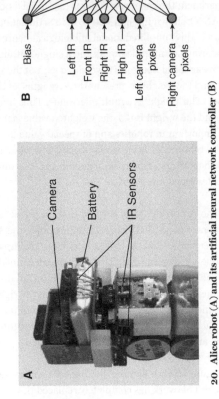

B

Left motor → Right motor → Share token

Bias

Left IR
Front IR
Right IR
High IR
Left camera pixels
Right camera pixels

20. **Alice robot (A) and its artificial neural network controller (B)**

microcontroller 'brain' of the robot. Here the robot has, in a slightly more literal sense, a 'central nervous system' that directly models — albeit in an ultra simplified way — biological nervous systems.

In Waibel, Floreano, and Keller's experiment, ten ALICE robots are foraging by finding and collecting physical tokens, then moving them to a 'nest' area. The experimental arena contained two kinds of token: small tokens that can be pushed by a single robot, and larger tokens too heavy for one robot to push on its own. The fitness of each robot was based on its success in foraging for tokens. In one experiment, the arena contained only large tokens, and robots successfully evolved the ability to cooperatively push these tokens to the nest.

However, when the arena contained both large and small tokens, the group 'kin structure', i.e. the genetic relatedness of the individuals, influenced the evolved behaviours. Groups of genetically unrelated robots evolved toward pushing the small tokens because this was the best way to maximize their own fitness. In contrast, related robots evolved altruistic behaviours — cooperating to push the large tokens at the expense of their own individual fitness. Remarkably, this experiment appears to confirm Hamilton's rule: that altruistic behaviours will evolve when the relatedness of individuals multiplied by the fitness benefit to the receiver of the altruistic behaviour is greater than the fitness cost of performing the behaviour.

Symbiotic evolutionary robot organisms

Imagine a group of robots sent into a collapsed building to search for survivors (fitted perhaps with the artificial whiskers of Chapter 3). The robots are small and mobile. Their small size is, of course, an advantage because they can access small spaces, and a large number of small mobile robots can search the building fast. But it's a collapsed building, which means that there are likely to be obstacles that these robots, because of their small size, cannot cross — perhaps gaps where part of the floor has fallen away, or fallen beams, or office cupboards, that block the robot's path and are too high for a single robot to climb over.

What if a robot, on encountering such a problem, were to be able to call for other robots nearby? When enough have gathered, these robots then physically join together to form a larger robot able to cross the gap or climb the high obstacle. On reaching the other side of the obstacle, the robots then disassemble, disperse, and continue their search as single robots.

A robot then finds a survivor but the ceiling above the survivor is fragile and likely to collapse. The robot again calls for other robots and they self-assemble into a different three-dimensional structure, this time to create a shell—a kind of strong robot shelter—over the survivor's head and upper body. They do this while also providing first aid to the survivor and vital audio and visual communication—together with location data—between the survivor and rescue workers outside the building. Thus the robot swarm provides physical aid and support to the survivor until rescue workers can safely extract them from the building.

This scenario may seem like something from science fiction, but it is not so far-fetched. A project called Symbrion is developing the fundamental robotics technologies that would be needed to realize this vision. Symbrion combines ideas from heterogeneous swarm robotics, modular or self-assembling robotics, evolutionary robotics, and strong elements of biological inspiration and modelling.

The aim of Symbrion is to create a heterogeneous swarm of robots that are individually highly capable (in terms of sensing and communication) and independently mobile, but that are able when required to autonomously self-assemble into three-dimensional 'organisms'. The exact shape and structure of a Symbrion organism would depend on the nature of the task faced by the Symbrion robots, and it would clearly be impossible to predict all possible shapes that might be needed, so the idea is that Symbrion would be able to adapt, in fact literally evolve, new body plans. Importantly the self-assembly process is reversible; when an organism is no longer needed it will autonomously disassemble back to its constituent individual robots. Individual robots thus alternate between *swarm mode*, in which they essentially operate in 2D, and *organism mode*.

Led by Paul Levi and Serge Kernbach, the Symbrion project team has created three types of basic robot, designed to (symbiotically) cooperate. They are called the *Backbone, Active wheel,* and *Scout* robots. Each robot is capable of independent locomotion and each also has a powered hinge joint. Each robot has a different type of locomotion and physical configuration, and hence different capabilities and strengths.

All three robots share a common docking interface, with docking ports on several faces, and each robot can therefore physically dock with at least two other robots. The docking interface provides a mechanically strong motor-actuated locking mechanism, in addition to electrical contacts that provide a power bus, together with wired network communication between docked robots.

From a biological point of view, it's both helpful and somewhat appropriate to think of the individual Symbrion robots as single cells and the Symbrion 'organism' as a multicellular organism. We have three 'species' of cells in one organism, but they are able to symbiotically cooperate and, for instance, share energy resources, or benefit from the massively increased motility of the aggregate multicellular organism.

The role of an individual Symbrion robot, or cell, within the organism will, of course, depend much on its physical position within that organism, and clearly the robot cells of the organism need to coordinate their actions so that the organism can, for instance, walk, climb, or bridge gaps. There is therefore a process analogous to cell differentiation: when the Symbrion organism self-assembles, individual robot cells will need to adopt a specialist function according to their position; when it self-disassembles, individual robot cells will in a sense revert to undifferentiated 'stem cell'-like robots.

Some robot cells might, for instance, find themselves as stiff structural elements, not required to move or actuate while in the organism; they act, in a sense, as bone cells. Other robot cells might find themselves acting as 'knee' cells, bending, when required, to create leg movements for a walking organism. The biological metaphor extends further. Since some robot cells in the organism will have greater energy demands than others—for instance, the knee cells will require more power than the bone cells—then the Symbrion organism will, via the power bus, provide not only energy sharing but energy homeostasis, so that an appropriate energy equilibrium is maintained across the whole organism.

Furthermore, we need to accept that individual robot cells will experience faults. Even with well engineered robots and organisms with small numbers of robot cells (i.e. less than twenty), the overall complexity of the organism means that faults will be inevitable. Symbrion organisms will therefore need an artificial immune system able to detect and in some way compensate for a fault in an individual robot cell.

Figure 21 (top) shows a 3D organism within the Symbrion simulator. Its body plan consists of seven robots to form an insect-like six-legged hexapod. The 'head' at the left is a Scout robot. The three pairs of 'legs' are Active wheel robots, each hinged to lift the whole organism off the ground. In between each Active wheel robot is a Backbone robot, and the 'tail' on the right is also a Backbone robot; it's as if the organism has three

Robotics

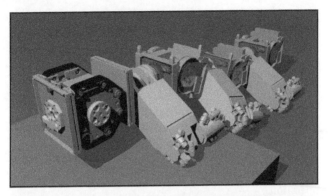

21. A Symbrion organism. Top: a visualization of the 3D organism. Bottom: Symbrion Backbone and Active wheel robots in the process of autonomous self-assembly

segments, each consisting of an Active wheel and Backbone pair. The hexapod is shown using its legs and backbone to climb onto a raised area, something that no individual robot could do on its own.

To form the organisms shown in Figure 21, individual Symbrion robots dock with each other to self-assemble in 2D, as planar body shapes, as shown in Figure 21 (bottom). Then the flattened organism coordinates the action of the various motors in the hinge joints of its component robot cells to lift itself off the ground and become a 3D organism.

The body plan of the Symbrion organism illustrated in Figure 21 was hand-designed. However, a goal of the project is to evolve body plans and to embed the process of artificial evolution within the robots themselves, a technique known as online evolution. The promise of this approach is that the robot collective will be able to evolve body plans, optimized to meet the particular needs of the tasks it has to solve, during operation.

Box 3 Robot simulators

In robotics research, the simulator has become an essential tool of the roboticist's trade. The reason for this is that designing, building, and testing successive versions of real robots is both expensive and time-consuming, and if part of that work can be undertaken in the virtual rather than the real world, development times can be shortened, and the chances of a robot that works first time substantially improved.

A robot simulator has three essential features. First, it must provide a virtual world. Second, it must offer a facility for creating a virtual model of the real robot. And third, it must allow the robot's controller to be installed and 'run' on the virtual robot in the virtual world; the controller then determines how the robot behaves when running in the simulator. The simulator should also provide a visualization of the virtual world and simulated robots in it so that the designer can see what's going on.

The first of these requirements, the virtual world, should be customizable. It must allow the robot developed to be able to create a reasonable facsimile of the robot's real operational environment, complete with the objects that the robot needs to sense, interact with, or physically manipulate. Importantly, the virtual world needs to simulate — with some reasonable degree of accuracy — the physics of the real world, so that when, for instance, solid objects collide, they don't simply pass through each other but react in a realistic way. Gravity also needs to be simulated, together with frictional forces, so that the movements of the virtual robot when accelerating or decelerating are realistic for the type of wheels or legs that the robot has and the surface that the robot is moving across.

To meet the second requirement, the simulator must provide a toolkit of models of real sensors and actuators. If the robot has, for instance, infrared proximity sensors, then the simulated sensors need to have — as far as possible — the same range and sensitivity as their real-world counterparts. Likewise the simulated motors must provide the same amount of movement in the virtual world as the real motors provide in the real world when given the same commands. A simulated camera in the virtual robot needs to be able to capture the same image that the real camera would see if mounted on the real robot in the same position and pointing in the same direction in the real world. Similarly, other sensors such as touch, proximity sensors, and laser-ranging sensors need to generate the same sense data as would be generated by the real sensors for the same obstacles in the same positions as in the real world. These are difficult challenges for developers of robot simulators.

Figure 22 shows a screen image of the simulator Webots simulating six humanoid Nao robots playing three-a-side football. Note that the small images show each of the robots' eye views of the field of play. This illustrates that simulators typically have built-in models for a range of existing robots, thus removing the

22. Simulated Nao humanoid robots playing 3-a-side football in the simulator Webots

need to create a virtual model of the real robot. Also illustrated here is the open competition Robostadium, played to the Standard Platform League rules for RoboCup, in which teams of robot programmers can develop and test algorithms for robot football, and complete with each other, all in simulation.

Chapter 6
Robotic futures

The challenge of writing a book like this is that its subject, the story of robotics, is still unfolding. Indeed, the pace at which robotics is advancing is accelerating, so there are likely to be significant developments in the near future, including some big surprises. By that I mean either currently overlooked corners of robotics research that turn out to be crucial, or robot technologies not yet invented that will have a huge impact.

Predicting these breakthroughs is impossible, so instead this chapter will address the question of robotic futures by outlining and discussing the technical problems that would need to be solved in order to build a number of 'thought experiment' robot systems: first, an autonomous planetary robot scientist; second, a swarm of medical micro-robots, and third, a humanoid robot companion with human-level intelligence.

Autonomous robot planetary scientist

Robots have a long and distinguished history in space exploration. Ever since Sputnik 1 was launched into earth orbit in 1957, space exploration has been dominated by unmanned spacecraft. Deep-space exploration has been exclusively undertaken by unmanned vehicles—think of the Voyager 1 and 2 spacecraft launched in 1977, remarkably still providing science data from the very edge of the solar system. It may seem odd to mention these vehicles in a book on robotics, but I would argue that they are in essence tele-operated robots. They are sensor-rich platforms with actuators, and a high degree of human remote control to make course corrections or to set up and operate the science instruments.

Less ambiguously robotic are the planetary rovers, which have given extraordinary service in the surface exploration of Mars, with the Mars exploration rovers Spirit and Opportunity famously far exceeding their planned mission lifetimes and science targets. There can be little doubt that robots will continue to lead the surface exploration of planetary bodies, including planetoids, moons, asteroids, and comets, perhaps paving the way for manned missions. But, as my brief introduction in Chapter 2 suggested, current planetary rovers represent the very limit of practical robot tele-operation. The inescapable physical limitation of communication delays significantly restricts the rover's speed of travel, which restricts the area that can be explored, and hence ultimately the science return from the mission. Given the very high cost of (even unmanned) planetary exploration, achieving more science for a given level of expenditure is clearly a significant driver, and one way to achieve more science is to increase the level of autonomy; in other words, build an autonomous planetary scientist.

Let's begin by thinking about the perfect robotic planetary scientist. It would, ideally, be sufficiently autonomous that, once delivered safely to the surface of the planet, it would carry out its mission to explore and survey, undertaking both geology and exobiology, calling home only to send status reports, images, and science data. The robot would be capable of deciding for itself how to explore its environment and which features are worthy of closer inspection. It would be able to plan and execute whatever physical moves are needed to get itself close to those features, then decide how best to use its science instruments to find out as much as it can about them. In short, it would need to behave like a human geologist and exobiologist rolled into one, noticing and investigating anything that its human counterparts would find interesting.

But, like a lone human explorer far from home, the robot would need to exercise caution. It would need an artificial sense of self-preservation so that it can decide when not to investigate a particular object closely because the risk to itself is judged too great (think of a tall rock formation that if disturbed might collapse onto the rover, or an object in a dangerous or inaccessible place, or simply something *very* unexpected).

Also like a human explorer, the ideal robot explorer would need to be adaptable and resourceful. If environmental conditions change, for instance, it should be able to adapt its activity to cope, perhaps 'going to sleep' to conserve energy. The rover should also be able to self-repair, so if parts of the robot become damaged, or fail through wear and tear, the robot can replace them with spares. Or if there are no spares, adapt its operation to compensate.

Of course, the science data that the rover collects is of great value, its very raison d'être, so if the robot should lose all communication and suffer catastrophic failure, it should be able to eject a data capsule, perhaps with a radio beacon, for a future rover to be able to find, rather like the explorer who perishes but leaves his diaries and notebooks to be found by those who come after.

What would the rover look like, and how would it move?

A robot planetary scientist with this level of capability and autonomy is far from the current state of the art, but is such a robot feasible, and what would be the major challenges? First consider the physical morphology of the robot and its means of locomotion. Much, of course, would depend upon how much is known of the nature of the planet's surface, its chemistry, temperatures, and pressures. If an atmosphere hides the planet's surface, then surface conditions have to be guessed at and, with good engineering and a lot of luck, the vehicle will survive long enough to send back useful data.

The Huygens probe, launched in 2004 from the Cassini spacecraft to the surface of Saturn's moon Titan, had a landing targeted onto a shoreline of unknown composition; the probe was designed for either solid landing or splashdown into a hydrocarbon ocean. But let us assume our target planet has solid ground to explore. In the development of Mars rovers, much work has been done on the design of locomotion systems, and a design with six powered wheels on rocker bogies is widely regarded as optimal for traversing rough rocky terrain. Shown in Figure 6, the rocker bogie suspension system enables the rover to safely drive over obstacles or through holes more than a wheel diameter (250 mm) in size.

The Mars rovers were designed to traverse desert regions with few large rocks (relative to the size of the rover), but what if the surface roughness of the planet is unknown? If the nature of the planet's surface is completely unknown but the planet has an atmosphere, then one option would be to fly above the surface, perhaps taking off, navigating, and landing between sites of interest. Flying vehicles for planetary exploration bring their own problems, but with a thick enough atmosphere there's no doubt that lighter-than-planetary-air vehicles present an attractive option.

Let's assume that either our planet doesn't have an atmosphere or it's too thin to support any kind of flying vehicle. Then what alternatives do we have? One option, particularly if gravity is low, perhaps on a planetoid or asteroid, is a robot capable of jumping. But this is again not without problems, because if you jump you don't know exactly where you are going to land, and it will be very difficult to guarantee not landing in a crevice. The safest option is a robot that carefully rolls or walks under its own control.

A radical approach to the physical design of the robot scientist would be a Symbrion-like structure in which the robot consists of a number of semi-independent self-assembling modules so that the robot can physically morph between different configurations, as the needs of the terrain require; for instance, a compact shape with wheels for fast travel on smooth terrain, or a spidery structure with long legs for a careful traverse across a boulder field.

One of the problems facing the designer of any planetary rover is that, even with excellent forward sensing, the robot cannot tell whether the ground it is about to roll onto is stable enough to take its weight. The Symbrion approach would again address this problem. Single modules could detach themselves, scout ahead, and literally probe the ground ahead to determine if it's safe, then return and reattach. If a scout module fails to return, the loss to the mission is not critical, since the robot scientist is an assembly of hundreds of modules.

How would the robot be scientist?

Let us turn now to the challenges presented by the robot's AI. As a combined geologist and exobiologist, it would need to be programmed with a large knowledge base of geology and biology. Given the capacity of mass storage, the idea that all that is known in these fields could be stored within the robot as a searchable database is entirely feasible.

Of course, being a good planetary scientist is not just about having a large number of facts. It's also about being able to ask the right question. Here the robot would undoubtedly benefit from an innovation in AI from the 1970s, namely the expert system: an approach for capturing the process by which a human expert exercises that expertise. One could imagine the robot planetary scientist going through a long training phase before launch, in which it is subject to a series of trials under the supervision of a human expert; so each time the robot makes a decision about what it 'believes' might be an interesting feature to investigate, the human expert either corrects errors or suggests alternatives, thus 'coaching' the robot.

If the robot's expert systems allow it to 'decide' what is interesting, providing it with artificial 'scientific curiosity', it would also need a system to moderate that curiosity with discretion—the robot's artificial 'self-preservation instinct'. The robot would, I suggest, need a system for self-assessing the risks against each action.

One possibility would be for the robot to have, like Cronos, a model of itself inside itself, in which it is able to set up and run simulations of each option available to it. This would be the equivalent of the robot being able to 'think through' possible courses of action and assign a risk to each one.

The robot's internal simulator would be preprogrammed with whatever mapping or survey data is available from orbiting satellites. Then, as the robot explores, it fills in the detail for those parts of the landscape that it has visited, or can scan or sense directly, and its internal model improves.

The robot's ability to come up with novel solutions to situations it finds itself in could be achieved by running a genetic algorithm, as described in Chapter 5, to evolve new behaviours. Or, if it's a Symbrion-type robot, evolve its own physical shape by testing a large number of variants of itself in the simulator.

The same approach could be extended to give the robot its self-repair characteristics. The robot has a model of itself inside the simulator, so it is able to search for ways to overcome the fault. If a particular wheel no longer functions, for instance, then the genetic algorithm can search for a new control behaviour or indeed physical morphology, to compensate for the fault.

Considered separately, none of these AI technologies is extreme or exotic, but what would be extremely challenging is putting them together and making them work. It should be noted also that the solutions outlined here for the design of an autonomous planetary scientist would apply equally to terrestrial robots, not only for autonomous exploration, surveying, or environmental monitoring, but rather more down-to-earth applications such as materials reclamation from landfill.

A swarm of medical micro-robots

Our second thought experiment might appear to be pure science fiction: a swarm of microscopic robots able to operate inside the human body and carry out medical procedures directly. The long-term vision would be a swarm that can be injected into the vascular system. The swarm of micro-robots would then literally swim to the source of the problem, either directed by a surgeon or with no external human control at all. Once they locate the problem, the micro-robots would ideally be able to signal their exact location and deliver therapeutics or undertake microsurgery directly.

Such a technology could revolutionize medicine and, more broadly, biology and biochemistry. The ability to robotically manipulate at the level of individual cells or even molecules would open up remarkable possibilities for new science. But is this vision a fantasy, or something that might realistically be developed in the medium-term future (i.e. several decades)? And if it is feasible, what are the technical problems that need to be solved?

The biggest challenge is, of course, miniaturization. To get an idea of how tough this problem is, a micro-robot able to access the smallest capillaries in the vascular system would need to be no bigger than 2 micrometres—2 millionths of a metre across. The smallest conventional robots we can fabricate are between 1,000 and 10,000 times too big. But remarkable progress has been made toward the goal of shrinking robots. Let me outline that progress.

Centibots

Miniature robots fall into three broad categories. Starting with the largest, these are robots that measure a few cubic centimetres. Let's call these *centibots*. These are now relatively commonplace in research labs. One example is the ALICE robot; the basic ALICE robot without the plug-ins shown in Figure 20 is little bigger than a sugar cube. Another example is the low-cost Kilobot, designed at Harvard University for swarm robotics experiments with perhaps thousands of robots. The body of each Kilobot is disk shaped, about 3 cm in diameter and 1 cm thick, and stands on three stiff wire legs; vibrating the legs causes the robot to move.

ALICE and Kilobot are both examples of autonomous robots that are assemblies of conventional miniature components (motors, electronics, sensors, and batteries). These centibots are, of course, far too large for the in-vivo medical applications in our thought experiment.

Millibots

The next size reduction requires design and fabrication techniques developed from chip design. The technology is called micro elecromechanical systems (MEMS), and robots based on this technology are called MEMS-based micro-robots. MEMS essentially uses the same photolithographic technology used to make integrated circuits (microchips), but mechanical (i.e. moving) components are fabricated together with electronic components, on the same chip. Although the individual components on MEMS-based micro-robots might be measured in micrometres (millionths of a metre), the whole robot will measure several millimetres. We could call these robots *millibots*.

Designing a complete working millibot based on MEMS technology is very challenging. At the time of writing, no one has succeeded in building and demonstrating a large-scale swarm of millibots, but a recent European project called I-SWARM went a long way toward achieving this goal. The prototype I-SWARM robots measured 3.9 mm × 3.9 mm × 3.3 mm. Standing on three legs, two at the front and one behind, with a single touch-sensing 'antenna' extending from the front, the robots might easily be mistaken for small insects.

The I-SWARM robots integrate all of the essential subsystems for swarm robotics on a flat 2D surface. Power is generated by a tiny solar cell on the top of the robot, which means that the experimental arena needs to be brightly lit. Robots communicate with each other using an ingenious micro-optical module. Locomotion and touch sensing are integrated into a single MEMS device that also acts as the robot's chassis. Three of the legs are angled downward to raise the robot off the surface; the fourth is left pointing directly ahead and acts as the robot's antenna. The legs and antenna are all actuated with an electroactive polymer, causing them to vibrate. By controlling which leg (or combination) is vibrating, the robot can be driven forward or back, or turned. The robot's antenna works by sensing the change in frequency of vibration when the antenna is touching something.

The I-SWARM robots are controlled by a conventional eight-bit microcomputer integrated with electronics for interfacing with the solar cell, communications module, legs, and antenna, all on a single custom application specific integrated circuit (ASIC). Another ingenious feature is the way the robots are programmed. Control code is downloaded to the robots by modulating the arena illumination.

Millibots, like the I-SWARM robots, are too big to be introduced into the vascular system (although robots of this size might conceivably be suitable for robotic heart surgery). Of course, any in-vivo application would need to solve several problems: how to swim, power, and encapsulation. Wireless pill cameras, which can be swallowed, are already being used to assist with diagnosis of problems in the digestive tract. It does seem likely that tele-operated pill-sized robots for medical interventions in the gut will become commonplace in the near future.

Microbots

The next big step in miniaturization, to true *micro-robots* — robots measuring a few micrometres — requires the solution of hugely difficult problems and, in all likelihood, the use of exotic approaches to design and fabrication. In order to get a sense of how big a step this is, consider that microbots need to be a thousand times smaller than millibots. Remarkably, however, work is already under way to address these challenges.

It is impossible to shrink mechanical and electrical components, or MEMS devices, in order to reduce total robot size to a few micrometres. In any event, the physics of locomotion through a fluid changes at the microscale and simply shrinking mechanical components from macro to micro — even if it were possible — would fail to address this problem. A radical approach is to leave behind conventional materials and components and move to a bioengineered approach in which natural bacteria are modified by adding artificial components. The result is a hybrid of artificial and natural (biological) components.

The bacterium has many desirable properties for a microbot. By selecting a bacterium with a flagellum, we have locomotion perfectly suited to the medium. A flagellum is a tiny 'tail' that is rotated by a remarkable molecular 'motor' built into the cell wall of the bacterium. Bacteria move and steer by switching this molecular motor on and off. Another hugely desirable characteristic is that the bacteria are able to naturally scavenge for energy, thus avoiding the otherwise serious problem of powering the microbots.

By engineering bacteria with additional magnetic materials, roboticists Sylvain Martel and Mahmood Mohammadi were able to 'steer' a single bacterium, with external electromagnets under computer control, and use it to push a 3 micrometre bead in the experimental fluid medium. In practice a single hybrid microbot would be useless. Instead we need a swarm, and in another astonishing experiment Martel and Mohammadi were able to create a single swarm of about 5,000 flagellated magnetotactic bacteria, again using external magnetic fields. In a demonstration of nanoscale assembly and manipulation, the swarm was then directed to move 'bricks', each 80 micrometres long, and assemble a tiny pyramid.

But, remarkable as they are, these experiments illustrate just one approach. It is possible that hybrid bacteria do not represent the best approach to engineered micro-robots. Other equally exotic approaches may ultimately prove more fruitful: for instance, microbots assembled by even smaller molecular nanomachines, an idea suggested by Richard Feynman and developed by Eric Drexler in his book *Engines of Creation*.

Whatever technology is used to create the microbots, huge problems would have to be overcome before a swarm of medical microbots could become a practical reality. The first is technical: how do surgeons or medical technicians reliably control and monitor the swarm while it's working inside the body? Or, assuming we can give the microbots sufficient intelligence and autonomy (also a very difficult challenge), do we forgo precise control and human intervention altogether by giving the robots the swarm intelligence to be able to do the job, i.e. find the problem, fix it, then exit? In this case, the microbot swarm could simply be injected into the bloodstream and left to do its work, just as any drug: it would, in effect, be a smart drug that does surgery.

But these questions bring us to what would undoubtedly represent the greatest challenge: validating the swarm of medical microbots as effective, dependable, and above all safe, then gaining approval and public acceptance for its use. This is a challenge that highlights an interesting question for roboticists like me who work on robot safety. Do we treat the validation of the medical microbot swarm as an engineering problem, and attempt to apply the same kinds of methods we would use to validate safety-critical systems such as air traffic control systems? Or do we instead regard the medical microbot swarm as a drug and validate it with conventional and (by and large) trusted processes, including clinical trials, leading to approval and licensing for use? My suspicion is that we will need a new combination of both approaches.

A humanoid robot companion

For our final thought experiment, consider a humanoid robot companion. This is an important example because humanoid robots are where robots started, and in many ways a humanoid robot with human-level artificial intelligence stands as a dream of robotics. Many roboticists see such a robot as the ultimate imitation of life, and so it could be said to represent the goal of the 'grand project' of robotics.

What do I mean by a humanoid robot companion? This robot would be an artificial companion. It would be capable of acting to some extent as a servant, a butler perhaps. But rather like the butler personified by P. G. Wodehouse's Jeeves, or Andrew the robot played by Robin Williams in the movie *Bicentennial Man*, our butler robot must be capable of not only serving and supporting its human mistress with everyday physical tasks, as a valet, but at the same time as a provider of company: a source of entertainment, comfort, and conversation. An entity with whom one can interact, converse, or even develop a friendship.

Wilks, in the book *Close Engagements with Artificial Companions*, suggests that the ideal characteristics for an artificial companion for middle-aged or elderly adults (which he labels a senior companion) might be close to the specification for an ideal Victorian companion: 'politeness, discretion, knowing their place, dependence, emotions firmly under control, modesty, wit, cheerfulness, well-informed, diverting, looks are irrelevant but must be presentable, long-term relationship if possible and limited socialization between companions permitted off-duty'. This implies a robot that is conversationally and linguistically highly capable, yet not overly (artificially) emotional. It must be very sensitive to its mistress's needs and feelings, demonstrate a degree of empathy but with an appropriate degree of detachment, and above all be trustworthy.

Let us set aside the question of whether such a robot companion is desirable (in the broad moral sense of whether such robots would be a good thing for society) and confine ourselves to the question of whether it is technically feasible. And assuming it is, how far is such a robot from practical realization? What technical problems would need to be solved in order to get from where we are now in terms of robot technology to where we would need to be to build such a robot?

First, consider what the robot should look like. We're assuming as a starting premise it should be humanoid, but — as I reflected in Chapter 4 — the term humanoid covers a very wide range of physical characteristics. Should just parts of the robot, perhaps the head and its arms, be high fidelity in appearance? Or should the robot be entirely android so that to some approximation it resembles a person, like Hiroshi Ishiguro's actroid or geminoid robots?

These are not just questions of cosmetics but important considerations. The robot needs to have an appearance that is neither unsettling nor absurd, given its role as valet and companion: somewhere between android and cartoon-like, perhaps.

The robot's arms, and especially the hands, are likely to come into closest proximity with humans—when handing food or drink to its mistress, or assisting with dressing or washing—so very great care and attention need to be given to their appearance, softness, and compliance.

In general terms the robot needs to be small and light enough that it cannot do any real harm if it bumps into or even falls onto a human. There is some evidence that a robot should be smaller than its human mistress in order not to seem intimidating: perhaps the size of a small human adult, say about 1.5 m with a body mass of 30 kg, roughly the weight of a ten-year-old child perhaps. In addition, the robot must be soft and compliant, at least as soft and compliant as a similarly sized human.

How would your robot companion sense you?

Consider now the sensors that our robot companion would need to be equipped with. Given its function as a personal valet and companion, the robot must be able to sense its human mistress with considerable acuity. As a minimum the robot must, firstly, be able to sense where and how far away, relative to itself, its mistress is. The robot must be able to hear her and localize the source of her voice, turning its head if necessary to face her in response to a verbal request for attention. Second, the robot should be able to see and track its mistress's posture, including, importantly, her head orientation and position of her arms and hands in order to 'read' her body gestures. Third, the robot should be capable of seeing its mistress's face with sufficient resolution to be able to recognize her, and her facial expressions; and, since humans use eye gaze to direct attention, the robot must be capable of gaze tracking.

The sensor technology to meet these requirements with some degree of sophistication exists already. Indeed, the body- and voice-tracking technology has already been developed for the interactive video games market in the shape of the Microsoft Kinect™ device, and it's no surprise that roboticists have been quick to experiment with this low-cost technology.

A robot companion should also be able sense its human mistress's general health and well-being, so additional sensors to monitor basic life signs, such as temperature and breathing, would also be a good idea. Even though it is not a nurse robot, a capable robot companion should surely be able to sense if there is something seriously wrong with its mistress and call for help.

Of course, although our robot companion's senses need to be focused on its human mistress, the robot also needs to be able to sense its environment and the objects in it with sufficient fidelity to be able to safely move around and through it. To support its valet functions, the robot must be able to sense and recognize objects well enough to be able to handle them, including handing them to its mistress (food or drink), or co-handling objects with its mistress (clothes, for instance, while assisting with dressing or undressing).

How would your robot companion know you?

Let us now turn to what is undoubtedly the major technical challenge we face in designing our robot companion: its artificial intelligence. At the outset I specified the robot as requiring human-level intelligence and, of course, in order to have a conversational capability equivalent to a human butler/ companion (witty, cheerful, well-informed, and diverting) our robot would indeed need human-level intelligence or a convincing emulation. Unfortunately, setting out a plausible shopping list of software 'modules' in the same way as I have done for the robot's sensorium is simply not possible. We just don't understand the architecture of human intelligence well enough to be able to construct a reasonable artificial emulation.

We can split the robot's AI into two major sections: one for its object recognition and manipulation and the other for its conversational human–robot capability.

Object handling and manipulation

As I outlined in Chapter 2, robots can be programmed with a particular skill: for instance, pouring liquid into a cup or folding clothes. But it doesn't make sense to program a robot for every possible object-handling eventuality—there are simply too many. It would make much more sense for the robot to be able to learn, by imitation, from human demonstration of the task (refer to Chapter 3 for an outline of robot learning). The robot would then be trained in a large number of basic tasks and skills before it would leave the factory.

But its development wouldn't stop there. The robot would need to be able to continue to learn new skills during its operational lifetime. This is a new way of thinking about machines: robots would be manufactured, then go through a developmental phase (this is a new sub-discipline called developmental robotics) in which they learn their own bodies, then go though a training phase in which they are taught skills and capabilities, and then — as working robots — they continue to learn and develop throughout their operational lifetimes.

It is interesting to realize that robots, once trained, would be able to pass their skills to other same-model robots in a way that humans cannot. Thus it's likely that for a particular model of robot companion, only the first would need to go through the full training regime. Subsequent robots would simply be preloaded with the learned skills from the first (calibrated for slightly different motors, sensors, etc). This also raises the intriguing possibility that if a robot companion while in service is faced with a task it cannot complete, it may be able to search an online database and simply download the new capability.

Our robot companion needs to be significantly more advanced than the current state of the art in object manipulation. Perhaps surprisingly, its conversational capability is much closer to reality than its object-manipulation skills.

Toward human-level AI: chatbots

Let us consider instead what we can build now, and try to assess how far short of the goal of human-level AI this falls. First, conversational input and output: these require speech recognition and speech synthesis—both now reasonably mature technologies. With its ability to learn the speech patterns and habits of the human speaker and cope with natural speech, speech-recognition software can properly be considered as incorporating some specialized AI functions.

Conversational AI programs have been around almost since the beginning of AI. The first was Joseph Weizenbaum's famous psychotherapist program, Eliza, of 1966. Modern online *chatbots* are surprisingly diverting. Try, for instance, the online Artificial Linguistic Internet Computer Entity ALICE. An interesting conversational robot can be achieved today by integrating speech recognition, a chatbot program, speech synthesis, and a module to control the lips and facial expression of the robot, such that the robot appears to both speak and reflect what it is speaking in its facial expressions.

Two examples of such robots, developed by Hanson Robotics, include the robot Philip K. Dick, and Bina48. The latter is especially interesting since a living person, Bina Rothblatt, commissioned the robot head to be a conversational robot emulation of herself. Bina48 (or, to be accurate, the robot's conversational program) was then 'trained' by hours of conversation between the real Bina and the robot Bina. *New York Times* reporter Amy Harmon famously interviewed Bina48 in 2010. It's clear that Harmon found the interview frustrating but with 'rare (but invariably thrilling) moments of coherence'. Such a marriage of AI chatbot and robotics technology demonstrates huge promise.

But conversation between two humans is much more than a verbal exchange. Our robot companion needs to be able to read and interpret the non-verbal gestures, body language, facial expressions, and eye gaze of its mistress, and use those often subtle cues as additional input of equal or greater importance to her spoken words.

Many of these gestures and their meanings are particular to a regional language and culture, but many are unique to the individual and will have to be learned by the robot. An effective robot companion will need to get to know its mistress well, and to do this we need to go far beyond chatbot technology. This brings me to propose a significant breakthrough that will be needed to bring our robot companion closer to reality: artificial *theory of mind*.

Artificial theory of mind

Theory of mind is the term given by philosophers and psychologists to the ability to form a predictive model of others. With theory of mind, it is supposed, we are able to understand how others might behave in particular circumstances. We can empathize because theory of mind allows us to imagine ourselves in the situation of others and hence feel as they feel.

However, the idea of theory of mind is empirically weak; we believe that it, or something like it, ought to exist but we have only the sketchiest understanding of the neurological or cognitive processes that might constitute theory of mind. Intelligent robotics provides an interesting approach to theory of mind, because it allows us to ask the question 'How would we build artificial theory of mind in a robot?'

When I described the robot Cronos in Chapter 4, I explained that Cronos has an internal model of itself. The purpose of that internal model, a computer simulation, is to allow Cronos to learn how to control itself. But if a robot can have an internal model of itself, then in theory it can have a model of others, including humans. Now it has to be said that this is a very big step. In fact it's several.

We need to go first from a robot with an internal model of itself to a robot with internal models of other robots exactly like itself. Such a model would allow a robot to predict (by running its simulation) what the other robot might do next, and modify its model if the other robot actually does something different. The next step is to go from a robot with an internal model of conspecifics to a robot with an internal model of another entity. For a robot companion, that other entity must, of course, be a human.

Again this is a very big step. (Perhaps a useful intermediate stage would be robots with predictive models of simpler animals.) And, what's more, there's no guarantee that a robot with an internal model of one particular human would demonstrate that it 'knows' that person. The so-called 'simulation' theory of mind is just a theory.

Despite the extraordinary difficulty, I believe that an effective robot companion must have some theory of mind in order to be able to demonstrate believable empathy (or a workable analogue). A robot companion with artificial theory of mind would be a much more knowing and empathic conversationalist than anything that could be achieved with chatbot technology.

It is interesting to speculate on what a robot companion with artificial theory of mind would be like. It would, I think, be very convincing: the physicality of the robot's gestures and facial expressions, its ability to read body language and modulate its own behaviour and responses according to not only what its mistress says and how she says it but also its internal model for her, could be a powerful and attractive (even dangerous) combination.

Would our robot companion have human-level artificial intelligence? I would say no. I can see no reason why the robot would, for instance, be creative or intuitive, or capable of deceit, joy, or despair, or any number of qualities we associate with what it is to be human. Some of these are, of course, undesirable qualities in an ideal companion, but the downside is that it would probably therefore not be capable of being creatively witty or diverting. A little dull perhaps, but quite possibly a very effective robot companion.

The meaning of robotics

For me, robots and the practice of robotics are profoundly fascinating because robots embody three ideas: first, that robots can be *useful machines*; second, that robots can be *working models* of life, intelligence, evolution, or even culture; and third, that robots are an *imitation of life*. The three ideas respectively represent different perspectives on robotics—engineering, science, and philosophy—and I find it deeply satisfying to work on robots that might offer insights from all three perspectives.

Much of this book has rightly focused on the first perspective: how robots are used now and how they might be increasingly useful in the future, the challenges that need to be overcome, and the technologies that need to be invented, before robotics can fulfil its extraordinary potential. I have touched upon aspects of the second perspective also, but not dwelt too long because the extent to which robots — as working models — have contributed to scientific understanding remains limited. It is hard to find incontrovertible examples of new biology, for instance, that resulted from robot models of biological processes.

I would, however, argue that the *idea* of robots as working models in science started in earnest with W. Grey Walter and his 1950 robot tortoises. What better example than a neurophysiologist choosing to build robots to test ideas on brain connectivity? It remains a serious idea and one that deserves to be explored. I sometimes describe robots as an *embodied simulation* and, in the same way that simulation is now recognized as a serious tool for scientific investigation with an emerging science of simulation, so we should seek a principled approach to how robots are used as working models in science.

But let me conclude this book by reflecting on the third idea: the philosophical notion of robots as an imitation of life. It is this idea that has perhaps the greatest resonance in popular culture: think of how often we read news headlines of robots that think, learn, or feel. Such headlines feed our science fiction fantasies but we react also with a thrill of fear; fictional robot dystopias appear to become just that little bit closer to reality. But setting aside the interesting subject of robots in popular culture, does the philosophical notion have any serious merit? I believe it does, because robots that act *as if* they are in some sense alive raise deep ontological questions about what it is to be alive, what it is to self-consciously experience the world and have agency and autonomy in the world.

In this book I have not addressed questions such as whether robots can think, be self-aware, or even conscious. These are controversial scientific and philosophical questions and it would seem that until we can solve the puzzle of how (or if) an animal or human can be conscious, there is little point attempting to engineer a conscious robot. After all, if we don't know what consciousness is, how would we measure whether a robot has it or not?

But I would argue that robotics should not simply wait until these hard questions are resolved. Instead, robotics can, I think, be part of the solution. By building robots that demonstrate ever more convincing emulations of agency, adaptability, and self-awareness, we can, I believe, cast new light on these philosophical questions. For instance, in my sketch of a future humanoid robot companion, I suggested that it would need artificial theory of mind. Would such a robot be self-aware, even conscious? I would say emphatically no, but I contend that embodied artificial theory of mind could be an important first step toward both building artificial and understanding natural consciousness.

Of course, our current anthropomorphic and zoomorphic robots are crude simulacra at best, but as the practice of robotics

131

advances they will become less crude. At some point in the far future, arguments that designed artificial things cannot be conscious will become more difficult to sustain. Think of the fictional android robot Data from *Star Trek: The Next Generation*: a robot with whom you can play music, discuss philosophy, sustain a friendship over many years, and mourn when it ceases to exist. Is such a robot theoretically possible?

Yes, I believe there is no reason *in principle* that humanoid robots could not be designed, or evolved, to behave as if they are conscious so convincingly that, for all intents and purposes, they should be regarded as actually being conscious. Such a perfect imitation might of course resolutely argue with a human philosopher that it is just a robot and not truly conscious, and the philosopher may well agree. But by any normal (enlightened) human standards, such a robot would also be capable of earning trust, respect, or even personhood. Such a robot would surely challenge our deeply held notions of what it is to be human.

Glossary

Technical terms used but not defined within this book.

3D printing: a technology for fabricating solid 3 dimensional parts, usually by injecting hot molten plastic from a nozzle. The nozzle is moved under computer control and the 3 dimensional part is built up from successive layers of printed plastic.

Actuator: in robotics this is the general term for any motorized part of a robot, i.e. mechanism used by the robot to act in its environment. Thus actuation means making the robot move or, literally, act.

Artificial Intelligence (AI): the science and engineering of intelligent machines, including computers. Robotics is thus the branch of AI concerned with physically embodied AI.

Chatbot: a computer programme designed to simulate a natural language conversation with a human. A chatbot is an example of an AI programme.

Collective Robotics: the general term for a system of multiple robots that work together as a team, group, or swarm to achieve a task.

Control System: the devices, usually electronics and software, which control a machine so that it achieves its required functionality.

End effector: the device (i.e. tool or gripper) fitted to the end of a robot arm.

Hall-effect sensor: a sensor that varies its output in response to a magnetic field.

Infra-red sensor: a short-range sensor that uses reflected infra-red (IR) light to measure the distance between the sensor and an object.

Laser Scanner: in robotics a laser range finder, a type of laser scanner, is used to create a large set of distance measurements from the scanner to objects in the robot's environment, by measuring the time-of-flight of each reflected pulse of laser light. Also known as Light Detection and Ranging (LIDAR).

Proximity sensor: the general name for a sensor whose function is to detect the presence of nearby objects, without physical contact. An infra-red sensor is an example of a proximity sensor.

Mechatronics: a term that refers to a combined system of mechanical, electronics and software.

Microcontroller: a computer (microprocessor) together with memory and input-output interfaces, integrated into a single device, used for embedded systems.

Sensor: a device that detects or measures a physical quantity and converts it to an electrical signal.

Servomotor: a motor with a built-in feedback mechanism that allows the output shaft of the motor to be positioned accurately.

Self-organization: when individuals follow local behavioural rules and this results in organized behaviour by the whole group, without the need for global control, the system is said to be self-organized.

Tele-presence: a robot whose primary function is to give its human user, who is remotely tele-operating it, some sense of being present at the robot's location is known as a tele-presence robot; also sometimes called an avatar.

Further reading

General introductions to robotics

Maja Matarić: *The Robotics Primer*, MIT Press, 2007. This book is a highly engaging and accessible semi-technical introduction to robotics, spanning all of the important concepts and applications.

Roland Siegwart, Illah R. Nourbakhsh, and Davide Scaramuzza: *Introduction to Autonomous Mobile Robots*, MIT Press, 2011. This is a comprehensive textbook on mobile robotics, ideal for anyone who wishes to understand robot motion, perception, and localization in greater technical depth.

Chapter 1: What is a robot?

Mark Elling Roshiem: *Leonardo's Lost Robots*, Springer, 2006. This beautifully illustrated book tells the extraordinary story of the reconstruction of Leonardo da Vinci's robots, including his programmable cart.

Philip Husbands, Owen Holland, and Michael Wheeler (eds): *The Mechanical Mind in History*, MIT Press, 2008. This is a fascinating and important account of the pioneers of intelligent machines, their work and ideas. It tells the story of how key ideas of bio-inspiration, which I introduce in Chapter 3, were conceived by the mid 20th-century pioneers of cybernetics.

Chapter 3: Biological robotics

Valentino Braitenberg: *Vehicles: Experiments in Synthetic Psychology*, MIT Press, 1984. A delightful book; the 'experiments' of the title are thought experiments of hypothetical autonomous vehicles.

Rodney A. Brooks: *Cambrian Intelligence: the Early History of the New AI*, MIT Press, 1999. This is a collection of Brooks's influential and very readable research papers from 1985 to 1991, with commentaries.

Chapter 4: Humanoid and android robots

Yoseph Bar-Cohen and David Hanson: *The Coming Robot Revolution: Expectations and Fears About Emerging Intelligent, Humanlike Machines*, Springer, 2009. A thought-provoking book which explores the larger societal implications of humanoid robotics.

Wendell Wallach and Colin Allen: *Moral Machines: Teaching Robots Right from Wrong*, Oxford University Press, 2008. This book provides a thorough and philosophically sound exploration of how we might build future ethical robots.

Endnote for section on Robot Ethics:

The set of five ethical principles for robotics outlined in this section were drafted by a joint Engineering and Physical Sciences (EPSRC) and Arts and Humanities Research Council (AHRC) working group on robot ethics, in 2010. The full report from that working group can be found at: http://www.epsrc.ac.uk/ourportfolio/ themes/engineering/activities/Pages/principlesofrobotics.aspx

Chapter 5: Robot swarms, evolution, and symbiosis

Eric Bonabeau and Guy Théraulaz: *Swarm smarts, Scientific American*, March 2000. A very readable review of swarm intelligence.

Dario Floreano and Lawrent Keller: 'Evolution of Adaptive Behaviour in Robots by Means of Darwinian Selection', *PLoS Biology*, Vol. 8, 2010. A very accessible article on evolving adaptive robot behaviours.

Chapter 6: Robotic futures

Paolo Dario and Arianna Menciassi: 'Robot Pills', *Scientific American*, August 2010. For an up-to-date review of research in pill-sized medical robots.

K. Eric Drexler: *Engines of Creation: the Coming Era of Nanotechnology*, Oxford University Press, 1990. Now regarded as a classic, this is a highly readable exploration of how nano-machines might be built and applied.

And beyond

Dario Floreano and Claudio Mattiussi: *Bio-inspired Artificial Intelligence*, MIT Press, 2008. A substantial textbook, but also a readable and comprehensive introduction to bio-inspired approaches to artificial intelligence and robotics; for anyone who wants a deeper understanding of these important new approaches this book is a must.

Rolf Pfeifer and Josh Bongard: *How the Body Shapes the Way We Think: A New View of Intelligence*, MIT Press, 2006. This clear and thoughtful book sets out the case for 'embodiment': that intelligence, in animals and robots, always requires a body. Its conclusions go well beyond robotics.

David McFarland: *Guilty Robots, Happy Dogs*, Oxford University Press, 2008. Written by an influential zoologist, this book is concerned with philosophy of mind. Written in an accessible and refreshing style, it is a very nice introduction to machine consciousness.

Online resources

Almost all of the robots and projects outlined in this book can be found on the Internet via a quick search. I also recommend:

Robots podcast: http://www.robotspodcast.com/
Interviews and background information, including video clips, with many of the world's leading roboticists.

Automation blog: http://spectrum.ieee.org/blog/robotics/robotics-software/automaton
An award-winning robotics blog from the IEEE.

The robot report: http://www.therobotreport.com/
A business-oriented website with news and links to robotics companies and research organizations.

Robotics News Service: http://www.hizook.com/

The author's blog: http://alanwinfield.blogspot.com/
Which includes a web page with additional links and resources for this book.

Index

INFORMATION
A Very Short Introduction
Luciano Floridi

Luciano Floridi, a philosopher of information, cuts across many subjects, from a brief look at the mathematical roots of information - its definition and measurement in 'bits'- to its role in genetics (we are information), and its social meaning and value. He ends by considering the ethics of information, including issues of ownership, privacy, and accessibility; copyright and open source. For those unfamiliar with its precise meaning and wide applicability as a philosophical concept, 'information' may seem a bland or mundane topic. Those who have studied some science or philosophy or sociology will already be aware of its centrality and richness. But for all readers, whether from the humanities or sciences, Floridi gives a fascinating and inspirational introduction to this most fundamental of ideas.

'Splendidly pellucid.'

Steven Poole, The Guardian

www.oup.com/vsi

INNOVATION
A Very Short Introduction
Mark Dodgson & David Gann

This *Very Short Introduction* looks at what innovation is and why it affects us so profoundly. It examines how it occurs, who stimulates it, how it is pursued, and what its outcomes are, both positive and negative. Innovation is hugely challenging and failure is common, yet it is essential to our social and economic progress. Mark Dodgson and David Gann consider the extent to which our understanding of innovation developed over the past century and how it might be used to interpret the global economy we all face in the future.

'Innovation has always been fundamental to leadership, be it in the public or private arena. This insightful book teaches lessons from the successes of the past, and spotlights the challenges and the opportunities for innovation as we move from the industrial age to the knowledge economy.'

Sanford, Senior Vice President, IBM

www.oup.com/vsi

Expand your collection of
VERY SHORT INTRODUCTIONS